Weakly Differentiable Mappings between Manifolds

of the
American Mathematical Society

Number 899

Weakly Differentiable Mappings between Manifolds

Piotr Hajłasz
Tadeusz Iwaniec
Jan Malý
Jani Onninen

March 2008 • Volume 192 • Number 899 (fourth of 5 numbers) • ISSN 0065-9266

American Mathematical Society
Providence, Rhode Island

2000 *Mathematics Subject Classification.* Primary 58D15; Secondary 46E35.

Library of Congress Cataloging-in-Publication Data

Weakly differentiable mappings between manifolds / Piotr Hajlasz...[et al.]
    p. cm. — (Memoirs of the American Mathematical Society, ISSN 0065-9266 ; no. 899)
Includes bibliographical references.
ISBN 978-0-8218-4079-5 (alk. paper)
1. Differentiable manifolds. 2. Sobolev spaces. I. Hajlasz, Piotr, 1966–

QA614.3.W43   2008
510 s—dc22
[516.3′6]
                                                      2007060584

Memoirs of the American Mathematical Society

This journal is devoted entirely to research in pure and applied mathematics.

Subscription information. The 2008 subscription begins with volume 191 and consists of six mailings, each containing one or more numbers. Subscription prices for 2008 are US$675 list, US$540 institutional member. A late charge of 10% of the subscription price will be imposed on orders received from nonmembers after January 1 of the subscription year. Subscribers outside the United States and India must pay a postage surcharge of US$38; subscribers in India must pay a postage surcharge of US$43. Expedited delivery to destinations in North America US$53; elsewhere US$130. Each number may be ordered separately; *please specify number* when ordering an individual number. For prices and titles of recently released numbers, see the New Publications sections of the *Notices of the American Mathematical Society*.

Back number information. For back issues see the *AMS Catalog of Publications*.

Subscriptions and orders should be addressed to the American Mathematical Society, P. O. Box 845904, Boston, MA 02284-5904, USA. *All orders must be accompanied by payment.* Other correspondence should be addressed to 201 Charles Street, Providence, RI 02904-2294, USA.

Copying and reprinting. Individual readers of this publication, and nonprofit libraries acting for them, are permitted to make fair use of the material, such as to copy a chapter for use in teaching or research. Permission is granted to quote brief passages from this publication in reviews, provided the customary acknowledgment of the source is given.

Republication, systematic copying, or multiple reproduction of any material in this publication is permitted only under license from the American Mathematical Society. Requests for such permission should be addressed to the Acquisitions Department, American Mathematical Society, 201 Charles Street, Providence, Rhode Island 02904-2294, USA. Requests can also be made by e-mail to reprint-permission@ams.org.

Memoirs of the American Mathematical Society is published bimonthly (each volume consisting usually of more than one number) by the American Mathematical Society at 201 Charles Street, Providence, RI 02904-2294, USA. Periodicals postage paid at Providence, RI. Postmaster: Send address changes to Memoirs, American Mathematical Society, 201 Charles Street, Providence, RI 02904-2294, USA.

© 2008 by the American Mathematical Society. All rights reserved.
Copyright of this publication reverts to the public domain 28 years
after publication. Contact the AMS for copyright status.
This publication is indexed in *Science Citation Index*®, *SciSearch*®, *Research Alert*®,
CompuMath Citation Index®, *Current Contents*®/*Physical, Chemical & Earth Sciences*.
Printed in the United States of America.

∞ The paper used in this book is acid-free and falls within the guidelines
established to ensure permanence and durability.
Visit the AMS home page at http://www.ams.org/

10 9 8 7 6 5 4 3 2 1   13 12 11 10 09 08

Contents

Chapter 1. Introduction   1

Chapter 2. Preliminaries Concerning Manifolds   7
- 2.1. Manifolds   7
- 2.2. The Sobolev space $\mathscr{W}^{1,p}(\mathbb{X},\mathbb{Y})$ 8
- 2.3. Differential forms   9
- 2.4. Mollifiers and smoothing operator   15
- 2.5. Maximal operators   16

Chapter 3. Examples   21
- 3.1. The longitude projection   21
- 3.2. Spherical coordinates   22
- 3.3. Winding around the longitude circles   22
- 3.4. A mapping of infinite degree   23

Chapter 4. Some Classes of Functions   25
- 4.1. Marcinkiewicz space $\mathscr{L}^p_{\text{weak}}(\mathbb{X})$ 25
- 4.2. The space $\mathscr{L}^{\alpha,p}(\mathbb{X})$ 26
- 4.3. The Orlicz space $\mathscr{L}^P(\mathbb{X})$ 28
- 4.4. Grand G$\mathscr{L}^p$-space   31
- 4.5. Relations between spaces   32
- 4.6. Sobolev classes   35

Chapter 5. Smooth Approximation   37
- 5.1. Web like structures   37
- 5.2. Vanishing web oscillations   37
- 5.3. Statements of the results   38
- 5.4. Proof of Theorem 5.1   39
- 5.5. Spinning a web on $\mathbb{X}$ 42
- 5.6. Proof of Theorems 1.1 and 1.2   44
- 5.7. Proof of Theorem 5.2   45
- 5.8. Proof of Theorem 1.3   45

Chapter 6. $\mathscr{L}^1$-Estimates of the Jacobian   47
- 6.1. Weak wedge products   48
- 6.2. Distributional Jacobian   49
- 6.3. Proof of Theorem 6.5   51

Chapter 7. $\mathscr{H}^1$-Estimates   55
- 7.1. The Hausdorff content   55
- 7.2. The $\mathscr{H}^1$-Theorem   56

Chapter 8. Degree Theory   63

8.1.	Definition of the degree via weak integrals	63
8.2.	Weak integrals	64
8.3.	Stability of the degree	66
8.4.	The degree in Orlicz and grand Sobolev spaces	67

Chapter 9. Mappings of Finite Distortion   69
   Acknowledgements   70

Bibliography   71

Abstract

We study Sobolev classes of weakly differentiable mappings $f : \mathbb{X} \to \mathbb{Y}$ between compact Riemannian manifolds without boundary. These mappings need not be continuous. They actually possess less regularity than the mappings in $\mathscr{W}^{1,n}(\mathbb{X}, \mathbb{Y})$, $n = \dim \mathbb{X}$. The central themes being discussed are:
- smooth approximation of those mappings
- integrability of the Jacobian determinant

The approximation problem in the category of Sobolev spaces between manifolds $\mathscr{W}^{1,p}(\mathbb{X}, \mathbb{Y})$, $1 \leqslant p \leqslant n$, has been recently settled in [2], [3], [17], [23], [24]. However, the point of our results is that we make no topological restrictions on manifolds $\mathbb{X}$ and $\mathbb{Y}$. We characterize, essentially all, classes of weakly differentiable mappings which satisfy the approximation property. The novelty of our approach is that we were able to detect tiny sets on which the mappings are continuous. These sets give rise to the so-called web like structure of $\mathbb{X}$ associated with the given mapping $f : \mathbb{X} \to \mathbb{Y}$.

The integrability theory of Jacobians in a manifold setting is really different than one might a priori expect based on the results in the Euclidean space. To our surprise, the case when the target manifold $\mathbb{Y}$ admits only trivial cohomology groups $H^\ell(\mathbb{Y})$, $1 \leqslant \ell < n = \dim \mathbb{Y}$, like n-sphere, is more difficult than the nontrivial case in which $\mathbb{Y}$ has at least one non-zero ℓ-cohomology. The necessity of topological constraints on the target manifold is a new phenomenon in the theory of Jacobians.

Received by the editor July 14, 2004.

2000 *Mathematics Subject Classification.* Primary 58D15; Secondary 46E35.

Key words and phrases. Sobolev spaces, mappings between manifolds, approximation, Jacobain, Hardy space, degree, rational homology sphere.

Hajłasz was supported by the KBN grant 2 PO3A 028 22 and also by the NSF grant DMS-0500966.

Iwaniec was supported by the NSF grants DMS-0301582 and DMS-0244297.

Malý was supported by the Research Projects MSM 113200007 and MSM 0021620839 from the Czech Ministry of Education, Grants No. 201/00/0767 and 201/03/0931 from the Grant Agency of the Czech Republic (GACR) and Grant No. 165/99 from the Grant Agency of Charles University (GAUK).

Onninen was supported by the National Science Foundation grant DMS-0400611.

CHAPTER 1

Introduction

Sobolev theory on Riemannian manifolds has come into widespread usage in modern geometry and topology. It also continues to be of great importance in nonlinear partial differential equations (PDE's for short), variational problems, like those in the theory of harmonic maps [26], [37] or quasiconformal deformations [32], [35], nonlinear elasticity, continuum mechanics, and much more. Looking ahead, we have attempted in this paper to present such mappings with all their nuances and possible applications.

The primary objects of our study are weakly differentiable mappings:

(1.1) $$f : \mathbb{X} \to \mathbb{Y}$$

where $\mathbb{X}$ and $\mathbb{Y}$ are smooth compact oriented Riemannian manifolds without boundary, $\dim \mathbb{X} = n \geqslant 2$ and $\dim \mathbb{Y} = m \geqslant 2$. One might say that C. B. Morrey [43] was the first to consider such mappings. The Sobolev class $\mathscr{W}^{1,p}(\mathbb{X}, \mathbb{Y})$ can be defined in a myriad of ways that are not always equivalent. In our approach we appeal to the celebrated theorem of J. Nash [47], which ensures that $\mathbb{Y}$ can be $\mathscr{C}^\infty$-isometrically imbedded in some Euclidean space $\mathbb{R}^N$. Let us assume that $\mathbb{Y} \subset \mathbb{R}^N$, for simplicity. This being so, we say that $f = (f^1, ..., f^N) : \mathbb{X} \to \mathbb{R}^N$ belongs to the Sobolev space $\mathscr{W}^{1,p}(\mathbb{X}, \mathbb{Y})$ if each coordinate function $f^i : \mathbb{X} \to \mathbb{R}$ lies in the usual Sobolev space $\mathscr{W}^{1,p}(\mathbb{X})$, and $f(x) \in \mathbb{Y}$ for almost every $x \in \mathbb{X}$. We do not reserve any particular notation of the Riemannian tensors on $\mathbb{X}$ and $\mathbb{Y}$, as these tensors will be fixed for the duration of this paper. The volume elements on $\mathbb{X}$ and $\mathbb{Y}$, denoted by $dx \in \mathscr{C}^\infty(\wedge^n \mathbb{X})$ and $dy \in \mathscr{C}^\infty(\wedge^m \mathbb{Y})$, will be the ones induced by the orientation and the metric tensors. In this way $\mathscr{W}^{1,p}(\mathbb{X}, \mathbb{Y})$, $1 \leqslant p < \infty$, becomes a complete metric subspace of the linear space $\mathscr{W}^{1,p}(\mathbb{X}, \mathbb{R}^N)$.

In the Riemannian manifolds setting it is not clear at all whether smooth mappings $f \in \mathscr{C}^\infty(\mathbb{X}, \mathbb{Y})$ are dense in $\mathscr{W}^{1,p}(\mathbb{X}, \mathbb{Y})$, a question raised by J. Eells and L. Lemaire [10]. This is trivially the case for $p > n$. R. Schoen and K. Uhlenbeck [49], [50] showed that the answer is also positive when $p = n$. That is all we can have in the category of the Sobolev spaces $\mathscr{W}^{1,p}(\mathbb{X}, \mathbb{Y})$, unless additional topological conditions are imposed on the manifolds $\mathbb{X}$ and $\mathbb{Y}$ [23], [24]. For example, in the same paper R. Schoen and K. Uhlenbeck [50] demonstrate that $\mathscr{C}^\infty(\mathbb{S}^n, \mathbb{S}^{n-1})$ is not dense in $\mathscr{W}^{1,p}(\mathbb{S}^n, \mathbb{S}^{n-1})$ for every $n-1 \leqslant p < n$. While it is not clear at this point, the Sobolev space $\mathscr{W}^{1,n}(\mathbb{X}, \mathbb{Y})$, with $n = \dim \mathbb{X} \geqslant 2$, will be the borderline case for many more phenomena concerning weakly differentiable mappings. Other related papers are [2], [3] [17], [18], [19], [20]. Sobolev spaces with exponents $1 < p < n$ are natural in the theory of harmonic mappings [26], [10], [37], [49] and other related areas. However, properties of these mappings are very different from those in $\mathscr{W}^{1,n}(\mathbb{X}, \mathbb{Y})$. This difference lies fairly deep in the concept of the topological degree. If $\dim \mathbb{X} = \dim \mathbb{Y} = n$, then a smooth mapping $f : \mathbb{X} \to \mathbb{Y}$ has well defined degree

$$\deg(f; \mathbb{X}, \mathbb{Y}) = \frac{1}{\operatorname{vol} \mathbb{Y}} \int_{\mathbb{X}} \mathcal{J}(x, f) \, dx \tag{1.2}$$

where $\mathcal{J}(x, f)$ stands for the Jacobian determinant of f. It is evident that this formula makes sense also for mappings in $\mathscr{W}^{1,n}(\mathbb{X}, \mathbb{Y})$. But it is less obvious whether it relates to topological properties of such mappings. Indeed it does, thanks to the density of $\mathscr{C}^{\infty}(\mathbb{X}, \mathbb{Y})$ in $\mathscr{W}^{1,n}(\mathbb{X}, \mathbb{Y})$. One might try to extend formula (1.2) to mappings $f \in \mathscr{W}^{1,p}(\mathbb{X}, \mathbb{Y})$, $1 < p < n$. For example, by assuming that the Jacobian is integrable. This attempt will fail, simply because there is no way to control the integral of the Jacobian by means of the p-norms of the differential of f. Actually, as f runs over $\mathscr{W}^{1,p}(\mathbb{X}, \mathbb{Y})$, $1 < p < n$, the integrals at (1.2) assume every real number.

In spite of these examples, we are still able to build a viable theory of weakly differentiable mappings slightly less regular than those in the Sobolev space $\mathscr{W}^{1,n}(\mathbb{X}, \mathbb{Y})$. One representative example is the Orlicz-Sobolev space $\mathscr{W}^{1,P}(\mathbb{X}, \mathbb{Y})$ of mapping $f : \mathbb{X} \to \mathbb{Y}$ whose differential $Df : \mathbf{T}\mathbb{X} \to \mathbf{T}\mathbb{Y}$ satisfies

$$\int_{\mathbb{X}} P(|Df(x)|) \, dx < \infty, \qquad P(t) = \frac{t^n}{\log(e+t)} \tag{1.3}$$

Let us emphasize, without getting into some technical details, that our theory will actually work for other Orlicz-Sobolev spaces $\mathscr{W}^{1,P}(\mathbb{X}, \mathbb{Y})$. But we must assume that P grows fast enough to satisfy the so-called *divergence condition*

$$\int_{1}^{\infty} \frac{P(t)}{t^{n+1}} \, dt = \infty \tag{1.4}$$

These classes, although appearing rather restrictive, contain $\mathscr{W}^{1,n}(\mathbb{X}, \mathbb{Y})$. However, they are typically smaller than the intersection of all the spaces $\mathscr{W}^{1,p}(\mathbb{X}, \mathbb{Y})$, $1 \leqslant p < n$.

$$\mathscr{W}^{1,n}(\mathbb{X}, \mathbb{Y}) \subset \mathscr{W}^{1,P}(\mathbb{X}, \mathbb{Y}) \subsetneq \bigcap_{1 \leqslant p < n} \mathscr{W}^{1,p}(\mathbb{X}, \mathbb{Y}) \tag{1.5}$$

We learn the necessity of the divergence condition (1.4) from the routine example of the radial projection

$$f_\circ : \mathbb{B}^n \to \mathbb{S}^{n-1}, \quad f_\circ(x) = \frac{x}{|x|} \tag{1.6}$$

As observed by R. Schoen and K. Uhlenbeck [50], $f_\circ$ cannot be approximated by smooth mapping $f_j : \mathbb{B}^n \to \mathbb{S}^{n-1}$ in the metric of $\mathscr{W}^{1,p}(\mathbb{B}^n, \mathbb{S}^{n-1})$ with any $p \geqslant n - 1$. This example, modified to manifolds without boundary, receives a thorough discussion in Section 3.1. Let us find out what we should assume on P to prevent $f_\circ$ from being a member of $\mathscr{W}^{1,P}(\mathbb{B}^n, \mathbb{S}^{n-1})$. The differential of $f_\circ$ belongs to the Marcinkiewicz space $\mathscr{L}^n_{\text{weak}}(\mathbb{B}^n)$. Precisely, we have $|Df_\circ(x)| = |x|^{-1}$ and hence

$$\int_{|Df_\circ| > t} dx = \frac{|\mathbb{B}^n|}{t^n}, \quad 1 \leqslant t < \infty \tag{1.7}$$

Integration in polar coordinates gives the formula

$$\int_{\mathbb{B}} P(|Df_\circ(x)|) \, dx = |\mathbb{S}^{n-1}| \int_{1}^{\infty} \frac{P(t)}{t^{n+1}} \, dt \tag{1.8}$$

Thus $f_\circ \notin \mathscr{W}^{1,P}(\mathbb{B}^n)$ if and only if P satisfies the divergence condition. That is why (1.3) is necessary to exclude $f_\circ$ from our theory.

In Section 5.6 we find the closure of $\mathscr{C}^\infty(\mathbb{X}, \mathbb{Y})$ in the Marcinkiewicz class $\mathscr{W}^{1,n}_{\text{weak}}(\mathbb{X}, \mathbb{Y})$.

Theorem 1.1. *The closure of $\mathscr{C}^\infty(\mathbb{X}, \mathbb{Y})$ in the metric topology of the Marcinkiewicz class $\mathscr{W}^{1,n}_{\text{weak}}(\mathbb{X}, \mathbb{Y})$ consists of mappings $f \in \mathscr{W}^{1,1}(\mathbb{X}, \mathbb{Y})$ such that*

$$(1.9) \qquad \lim_{t \to \infty} t^n \int_{|Df|>t} dx = 0$$

This condition is slightly stronger than that of requiring $|Df| \in \mathscr{L}^n_{\text{weak}}(\mathbb{X})$. It is worth noting that for every $0 \leqslant \alpha < n$, the condition (1.9) is equivalent to

$$(1.10) \qquad \lim_{t \to \infty} t^{n-\alpha} \int_{|Df|>t} |Df(x)|^\alpha \, dx = 0$$

In spirit similar to that of Theorem 1.1, we formulate our most general approximation result.

Theorem 1.2. *Every weakly differentiable mapping $f : \mathbb{X} \to \mathbb{Y}$, $\dim \mathbb{X} = n$, satisfying*

$$(1.11) \qquad \liminf_{t \to \infty} t^{n-p} \int_{|Df|>t} |Df(x)|^p \, dx = 0 \, , \qquad n-1 < p < n$$

can be approximated by smooth mappings $f_j : \mathbb{X} \to \mathbb{Y}$ in the metric topology of $\mathscr{W}^{1,p}(\mathbb{X}, \mathbb{Y})$.

This seemingly insignificant replacement of "lim" in (1.10) by "lim inf", has far reaching advantages. Among them is the density of $\mathscr{C}^\infty(\mathbb{X}, \mathbb{Y})$ in the Orlicz-Sobolev spaces $\mathscr{W}^{1,P}(\mathbb{X}, \mathbb{Y})$. In addition to (1.4), we shall impose some minor technical assumptions on P, see Theorem 5.2.

One further category of mappings appears in a natural way; the *grand Sobolev space*, denoted by $G\mathscr{W}^{1,n}(\mathbb{X}, \mathbb{Y})$. It consists of mappings

$$f \in \bigcap_{1 \leqslant s < n} \mathscr{W}^{1,s}(\mathbb{X}, \mathbb{Y})$$

such that

$$(1.12) \qquad \|Df\|_{n)} \stackrel{\text{def}}{=\!=} \sup_{0 < \epsilon \leqslant n-1} \left(\epsilon \int_{\mathbb{X}} |Df(x)|^{n-\epsilon} dx \right)^{\frac{1}{n-\epsilon}} < \infty$$

To illustrate, this space contains $\mathscr{W}^{1,n}_{\text{weak}}(\mathbb{X}, \mathbb{Y})$. Rather than discuss this space in full details, let us introduce a subspace $V\mathscr{W}^{1,n}(\mathbb{X}, \mathbb{Y}) \subset G\mathscr{W}^{1,n}(\mathbb{X}, \mathbb{Y})$ characterized by the condition,

$$(1.13) \qquad \lim_{\epsilon \to 0} \epsilon \int_{\mathbb{X}} |Df(x)|^{n-\epsilon} dx = 0$$

Our consideration of this subspace is motivated by the following result:

Theorem 1.3. *Smooth mappings are dense in $V\mathscr{W}^{1,n}(\mathbb{X}, \mathbb{Y})$.*

All known proofs of the density of smooth mappings in $\mathscr{W}^{1,n}(\mathbb{X}, \mathbb{Y})$ are based on the embedding of $\mathscr{W}^{1,n}(\mathbb{X}, \mathbb{Y})$ into $\text{VMO}(\mathbb{X}, \mathbb{Y})$ -the space of mappings of vanishing mean oscillation [6], [7]. It turns out that the spaces $\mathscr{W}^{1,P}(\mathbb{X}, \mathbb{Y})$ and $V\mathscr{W}^{1,n}(\mathbb{X}, \mathbb{Y})$, for which we prove density results, do not admit embeddings in $\text{VMO}(\mathbb{X}, \mathbb{Y})$. Thus we had to use a completely different idea. Our proofs of smooth approximation involve an interesting new device, the so-called web like structure on $\mathbb{X}$. For a somewhat related approximation method see [21]. A web on $\mathbb{X}$ is a compact set $\mathbb{F} \subset \mathbb{X}$ of measure zero whose complement consists of a finite number of components, disjoint open connected sets called meshes of the web.

Given $f : \mathbb{X} \to \mathbb{Y}$, as in Theorem 1.2, there exist webs $\mathbb{F} \subset \mathbb{X}$, with meshes as small as we wish, so that f restricted to $\mathbb{F}$ is continuous. Moreover, oscillations of f over the boundary of every mesh of the web can be made arbitrarily small. And that is why we say; f has vanishing web oscillations. The presence of small oscillations of mappings in $\mathscr{W}^{1,n}(\mathbb{X}, \mathbb{Y})$ or $\mathrm{V}\mathscr{W}^{1,n}(\mathbb{X}, \mathbb{Y})$, and in other Sobolev subclasses of $\mathscr{W}^{1,p}(\mathbb{X}, \mathbb{Y})$ with exponent p below the dimension of $\mathbb{X}$, seems to be important in future applications.

Now, that we have the approximation theorems, we will be able to give meaning to usually divergent integrals of the Jacobian of $f : \mathbb{X} \to \mathbb{Y}$, $\dim \mathbb{X} = \dim \mathbb{Y} = n$. Like in the Euclidean case [1], [45], [13], [32] this leads to a definition of the distributional Jacobian. In various situations the manifold setting is really different than one might a priori expect. Manifolds of the same deRham cohomology groups as $\mathbb{S}^n$ will be named *rational homology spheres*. This class of manifolds contains all homology spheres (manifolds whose integral homology groups are the same as those of the sphere) though these two classes are not the same. Indeed, for $p > 1$ the lens spaces $L(p,q)$ are 3-dimensional rational homology spheres, but their integral homology[1] groups are different from those of $\mathbb{S}^3$, see e.g. Proposition 21.28 in [15]. To our surprise the case when the target manifold $\mathbb{Y}$ is a rational homology sphere is more difficult than all other cases. What makes a difference is that in these other cases every n-form $\omega \in \mathscr{C}^\infty(\wedge^n \mathbb{Y})$ decomposes into wedge products of closed forms of degree smaller than n; that is,

$$(1.14) \qquad \omega = \sum_{i=1}^{K} \alpha_i \wedge \beta_i$$

where

$$(1.15) \qquad \begin{cases} \alpha_i \in \mathscr{C}^\infty(\wedge^{\ell_i} \mathbb{Y}) \cap \ker d, & \ell_i, k_i \in \{1, 2, ..., n-1\} \\ \beta_i \in \mathscr{C}^\infty(\wedge^{k_i} \mathbb{Y}) \cap \ker d, & \ell_i + k_i = n \end{cases}$$

Such is not the case of the n-sphere $\mathbb{Y} = \mathbb{S}^n$. These forms, once pulled back via a mapping $f : \mathbb{X} \to \mathbb{Y}$, bring us to analogous decomposition of $f^\sharp \omega$ in $\mathbb{X}$,

$$(1.16) \qquad f^\sharp \omega = \sum_{i=1}^{K} f^\sharp \alpha_i \wedge f^\sharp \beta_i$$

Under suitable regularity of the mapping f, the pullbacks $f^\sharp \alpha_i$ and $f^\sharp \beta_i$ are closed forms. At this point, a careful reader may observe that the dimension of $\mathbb{Y}$ is immaterial if we confine ourselves to pullbacks of the wedge products at (1.14), with $k_i + \ell_i = n \leqslant \dim \mathbb{Y}$. We refer to (1.14) as *Cartan forms*, named after H. Cartan who studies similar decompositions of differential forms. These ideas fit into even larger framework. The pullback at (1.16) is just a special case of a Cartan form on $\mathbb{X}$, the domain of f. Precisely, the n-form $\Lambda = f^\sharp \omega$ on $\mathbb{X}$ admits Cartan's decomposition as well:

$$(1.17) \qquad \Lambda = \sum_{i=1}^{K} \Phi_i \wedge \Psi_i$$

---

[1] In what follows by cohomology we will always mean deRham cohomology

where

(1.18) $$\begin{cases} \Phi_i \in \mathscr{L}^{p_i}(\wedge^{\ell_i}\mathbb{X}) \cap \ker d\,, & \ell_i, k_i \in \{1, 2, ..., n-1\} \\ \Psi_i \in \mathscr{L}^{r_i}(\wedge^{k_i}\mathbb{X}) \cap \ker d\,, & \ell_i + k_i = n \end{cases}$$

If we assume that $f \in \mathscr{W}^{1,s}(\mathbb{X}, \mathbb{Y})$ for some $s \geqslant n-1$, then $f^\sharp \alpha_i = \Phi_i \in \mathscr{L}^{\frac{s}{\ell_i}}(\wedge^{\ell_i}\mathbb{X})$ and $f^\sharp \beta_i = \Psi \in \mathscr{L}^{\frac{s}{k_i}}(\wedge^{k_i}\mathbb{X})$.

The integral $\int_\mathbb{X} \mathcal{J}(x, f)\, dx = \int_\mathbb{X} \Lambda$ exists in a well defined weak sense. It will actually converge if $\mathcal{J}(x, f) \geqslant 0$, because of Monotone Convergence Theorem. We refer to mappings having nonnegative Jacobian as *orientation preserving*, which not necessarily agrees with the term commonly used in topology.

The situation is dramatically different if one cannot decompose $f^\sharp \omega$ into wedge products. To illustrate, we give the following rather striking result.

THEOREM 1.4. *Every Orlicz-Sobolev class $\mathscr{W}^{1,P}(\mathbb{S}^n, \mathbb{S}^n)$, with $P(t) = o(t^n)$, contains an orientation preserving mapping whose Jacobian is not integrable.*

This contrasts sharply with the situation for mappings into $\mathbb{R}^n$, see [34] and [36].

Mapping supporting Theorem 1.4 are constructed in Section 3. This amounts to saying that if the target manifold $\mathbb{Y}$ is a rational homology sphere, then the classical integral formula for the degree of f fails in every Orlicz-Sobolev class below $\mathscr{W}^{1,n}(\mathbb{X}, \mathbb{Y})$.

It is evident from the Sobolev imbedding theorem that two mappings $f, g \in \mathscr{W}^{1,p}(\mathbb{X}, \mathbb{Y})$, with $p > n = \dim \mathbb{X}$, which are sufficiently close in $\mathscr{W}^{1,p}(\mathbb{X}, \mathbb{Y})$, are homotopic. Here $\mathbb{X}$ and $\mathbb{Y}$ may have different dimension and need not be even orientable. In [56] and [57], B. White proved a stronger result according to which every two continuous mappings of the Sobolev class $\mathscr{W}^{1,n}(\mathbb{X}, \mathbb{Y})$ are homotopic, provided they are sufficiently close in $\mathscr{W}^{1,n}(\mathbb{X}, \mathbb{Y})$. This and the fact that smooth mappings are dense in $\mathscr{W}^{1,n}(\mathbb{X}, \mathbb{Y})$ allow us to define homotopy classes in the space $\mathscr{W}^{1,n}(\mathbb{X}, \mathbb{Y})$.

The above reflections on the homotopy classes suggest several natural questions, like the following one:

QUESTION 1.5. *Do there exist any topological conditions on $\mathbb{X}$ and $\mathbb{Y}$ under which every two mappings $f, g \in \mathscr{W}^{1,p}(\mathbb{X}, \mathbb{Y}) \cap \mathscr{C}(\mathbb{X}, \mathbb{Y})$, $n-1 < p < n = \dim \mathbb{X}$, sufficiently close in the metric of $\mathscr{W}^{1,p}(\mathbb{X}, \mathbb{Y})$, are homotopic?*

This is not away the case for continuous mappings in $\mathscr{W}^{1,p}(\mathbb{X}, \mathbb{S}^n)$. However, the answer is in the affirmative if $\pi_n(\mathbb{Y}) = 0$, where we even do not require that $\dim \mathbb{Y} = n$. If, in addition, $\pi_{n-1}(\mathbb{Y}) = 0$ then smooth mappings are dense in $\mathscr{W}^{1,p}(\mathbb{X}, \mathbb{Y})$ by [24]. Note also that the homotopy condition is of a different nature than the cohomological one. Indeed, it is not difficult to construct a manifold $\mathbb{Y}$ of dimension n with nontrivial cohomology group $H^\ell(\mathbb{Y})$ for some $1 \leqslant \ell < n$ and such that $\pi_n(\mathbb{Y}) \neq 0$. This shows that continuity of the degree can not be deduced from the homotopy equivalence of mappings under the assumption that $\pi_n(\mathbb{Y}) = 0$.

The difference between trivial and nontrivial target space in terms of its ℓ-cohomologies $1 \leqslant \ell < n$, can also be detected in the borderline Sobolev space $\mathscr{W}^{1,n}(\mathbb{X}, \mathbb{Y})$. Two well known results in $\mathbb{R}^n$ are worthwhile to consider also on manifolds. Our first result asserts that if the Jacobian determinant $\mathcal{J}(x, f)$ of a mapping $f \in \mathscr{W}^{1,n}(\mathbb{X}, \mathbb{Y})$ is a priori nonnegative then $\mathcal{J}(x, f)$ belongs to the Zygmund space $\mathscr{L} \log \mathscr{L}(\mathbb{X})$. The second result asserts that if, in addition, $\mathbb{Y}$ is not a rational homology sphere, then we also have a uniform bound,

$$\int_{\mathbb{X}} \mathcal{J}(x,f) \log\left(e + \frac{\mathcal{J}(x,f)}{\int_{\mathbb{X}} \mathcal{J}(z,f)\,dz}\right) dx \preccurlyeq \int_{\mathbb{X}} |Df|^n \tag{1.19}$$

Remark. Throughout this paper we use the symbol $\preccurlyeq$ to indicate that the inequality holds with certain positive constant in the right hand side. This constant, referred to as *implied constant*, will vary from line to line. In all instances the reader may easily recognize which parameters the implied constant depends on. If not, we will explicitly specify those parameters. For example in (1.19), we will inform the reader that the implied constant depends only on the manifolds $\mathbb{X}$ and $\mathbb{Y}$.

The $\mathscr{L} \log \mathscr{L}$-integrability of $\mathcal{J}(x,f)$ remains true in case of the rational homology sphere space but the arguments will be completely different. Unexpectedly, the uniform bound (1.19) is lost. If the Jacobian changes sign then it still belongs to the Hardy space $\mathscr{H}^1(\mathbb{X})$, a well known result by R. Coifman, P. Lions, Y. Meyer and S. Semmes [9] in $\mathbb{R}^n$, see also [33], [36]. Again, in manifold setting the arguments establishing $\mathscr{H}^1$-regularity of the Jacobian will be more subtle than in $\mathbb{R}^n$. We have a uniform bounds only when the target manifold has a nontrivial cohomology; that is $H^\ell(\mathbb{Y}) \neq 0$ for some $1 \leqslant \ell < n$. Precisely, the estimate takes the form

$$\|f^\sharp \omega\|_{\mathscr{H}^1(\mathbb{X})} \preccurlyeq \int_{\mathbb{X}} |Df|^n \tag{1.20}$$

These and many more new results will be discussed at full length throughout this work. But first some background material is in order.

CHAPTER 2

Preliminaries Concerning Manifolds

This section is written to provide notation and to serve as brief introduction to the $\mathscr{L}^p$-theory of differential forms. The general references here are [8], [43], [51], [27] and [35], [55].

2.1. Manifolds

While many geometric constructions in $\mathbb{R}^n$ can be transferred to the Riemannian manifolds, the sometimes cumbersome technical details are often new and desired. Many unfamiliar differences will be explicitly emphasized here. Those differences sometimes only technical, sometimes delicate and important, are scattered throughout the research journals. Although, most of these facts will be left unproven in this text, we state them clearly so that they are available for a routine verification.

Our ambient space, subject to weakly differentiable deformations, will be an oriented compact (without boundary) smooth Riemannian manifold $\mathbb{X}$ of dimension $n \geqslant 2$.

2.1.1. Legitimate balls. Making precise estimates demands that we must work with one atlas $\mathcal{A}$ consisting of a finite number of coordinate charts $(\Omega, \kappa) \in \mathcal{A}$, where $\kappa : \Omega \to \mathbb{R}^n$ is a $\mathscr{C}^\infty$-diffeomorphism of an open region $\Omega \subset \mathbb{X}$ *onto* $\mathbb{R}^n$. Let us choose and fix such an atlas $\mathcal{A}$ and call it the reference atlas. We then introduce the so-called *reliable* radius of the manifold $\mathbb{X}$. This is a positive number, denoted by $R = R_\mathbb{X}$, such that for $0 < r \leqslant R_\mathbb{X}$ every pair of concentric geodesic balls $\mathbb{B}(x,r) \subset \mathbb{B}(x,4r) \subset \mathbb{X}$ fits in one coordinate region Ω of the atlas $\mathcal{A}$. We refer to such $\mathbb{B}(x,r)$ as *legitimate* ball in $\mathbb{X}$. The point to introducing this concept is that estimates on a legitimate ball can be reduced equivalently to analogous estimates in the Euclidean space. We mention now that the legitimate balls $\mathbb{B} = \mathbb{B}(x,r) \subset \mathbb{X}$ share basic properties of the Euclidean balls. In particular,

(2.1) $$|\mathbb{B}| \preccurlyeq (\operatorname{diam} \mathbb{B})^n \preccurlyeq |\mathbb{B}|$$

Here the implied constant depends only on the manifold $\mathbb{X}$.

2.1.2. Whitney covering. The familiar decomposition of an open set $\Omega \subset \mathbb{R}^n$ into Whitney cubes can be adapted to manifolds. While cubes are perfect regions for constructing various partitions of $\mathbb{R}^n$, there are serious geometric and combinatorial obstacles to do the same on manifolds. We shall work with the legitimate balls instead of cubes. Since it is impossible to partition a manifold into mutually disjoint balls, we will work with a finite covering in which the number of overlapping balls depends only on the manifold $\mathbb{X}$.

PROPOSITION 2.1. *Given a non-empty open set $\Omega \subsetneq \mathbb{X}$ and its complement $\mathbb{F} = \mathbb{X} \setminus \Omega$. There exists a collection $\mathcal{F} = \{\mathbb{B}_1, \mathbb{B}_2, ...\}$ of legitimate balls $\mathbb{B}_i \subset \mathbb{X}$ such that*

1) $\mathbb{B}_i \subset 2\mathbb{B}_i \subset \Omega$, $i = 1, 2, \ldots$
2) $\Omega = \bigcup_{i=1}^{\infty} \mathbb{B}_i$
3) $\sum_{i=1}^{\infty} \chi_{2\mathbb{B}_i}(x) \preccurlyeq 1$ for all $x \in \Omega$
4) $\operatorname{diam} \mathbb{B}_i \preccurlyeq \operatorname{dist}(\mathbb{B}_i, \mathbb{F}) \preccurlyeq \operatorname{diam} \mathbb{B}_i$ for all $i = 1, 2 \ldots$

Hereafter, $\chi_{\mathbb{E}}$ denotes the characteristic function of a measurable set $\mathbb{E} \subset \mathbb{X}$. Also, $2\mathbb{B}$ stands for the ball of the same center as $\mathbb{B}$ but with radius 2 times larger.

2.2. The Sobolev space $\mathscr{W}^{1,p}(\mathbb{X}, \mathbb{Y})$

The various classes of mappings $f : \mathbb{X} \to \mathbb{Y}$ in this paper are defined based on the classical Sobolev theory of real valued functions. Note that $\mathscr{W}^{1,p}(\mathbb{X}) = \mathscr{W}^{1,p}(\mathbb{X}, \mathbb{R})$ is a Banach space equipped with the norm

$$\|f\|_{\mathscr{W}^{1,p}} = \|f\|_{\mathscr{L}^1} + \|Df\|_{\mathscr{L}^p} \tag{2.2}$$

We adopt the classical results in $\mathbb{R}^n$ to our manifold setting, see for instance [48].

LEMMA 2.2. *Smooth functions in $\mathscr{C}^{\infty}(\mathbb{X})$ are dense in $\mathscr{W}^{1,p}(\mathbb{X})$, $1 \leqslant p < \infty$.*

LEMMA 2.3. [POINCARÉ INEQUALITY] *For every set $\mathbb{E}$ of a positive measure in $\mathbb{X}$ and $f \in \mathscr{W}^{1,p}(\mathbb{X})$ we have*

$$\int_{\mathbb{X}} |f - f_{\mathbb{E}}|^p \leqslant C_{\mathbb{E}} \int_{\mathbb{X}} |Df(x)|^p \, dx$$

The constant $C_{\mathbb{E}}$ actually depends only on the measure of $\mathbb{E}$. As usual, the integral average of f over the set $\mathbb{E}$ is denoted by

$$f_{\mathbb{E}} = \fint_{\mathbb{E}} f(x) \, dx = \frac{1}{|\mathbb{E}|} \int_{\mathbb{E}} f(x) \, dx$$

The local variant of Poincaré inequality reads as:

LEMMA 2.4. *For every legitimate ball $\mathbb{B} = \mathbb{B}(a, r)$, we have*

$$\int_{\mathbb{B}} |f - f_{\mathbb{B}}|^p \preccurlyeq r^p \int_{\mathbb{B}} |Df(x)|^p \, dx, \quad \text{whenever } f \in \mathscr{W}^{1,p}(\mathbb{X})$$

As a matter of fact this inequality is true for all geodesic balls in $\mathbb{X}$, but we shall exploit this inequality only for legitimate balls. Regarding the implied constant, we must emphasize that it depends neither on f nor on the radius r.

Now given two Riemannian manifolds $\mathbb{X}$ and $\mathbb{Y}$, we shall consider the Sobolev space $\mathscr{W}^{1,p}(\mathbb{X}, \mathbb{Y})$ of mappings whose tangent linear map (differential)

$$Df(x) : T_x \mathbb{X} \to T_y \mathbb{Y}, \quad y = f(x) \tag{2.3}$$

is $\mathscr{L}^p$-integrable. Our description, and certainly a rigorous definition of $\mathscr{W}^{1,p}(\mathbb{X}, \mathbb{Y})$, relies on an imbedding $\mathbb{Y} \subset \mathbb{R}^N$ [47].

THEOREM 2.5. *(J. Nash) Every $\mathscr{C}^{\infty}$-smooth Riemannian manifold $\mathbb{Y}$ can be $\mathscr{C}^{\infty}$-isometrically imbedded in some Euclidean space $\mathbb{R}^N$.*

The reader is also referred to M. L. Gromov and V. A. Rohlin [16] for an account of the imbedding problem. The Nash theorem allows us to consider $\mathscr{W}^{1,p}(\mathbb{X}, \mathbb{Y})$ as a subclass of a linear space of mappings $f : \mathbb{X} \to \mathbb{R}^N$ such that $f(x) \in \mathbb{Y}$ at almost every $x \in \mathbb{X}$. The metric topology in $\mathscr{W}^{1,p}(\mathbb{X}, \mathbb{Y})$ will be inherited from the associated norm topology in the linear space $\mathscr{W}^{1,p}(\mathbb{X}, \mathbb{R}^N)$. In this way the Sobolev class $\mathscr{W}^{1,p}(\mathbb{X}, \mathbb{Y})$ becomes a complete metric space. In what follows we shall tacitly use the fact that $\mathscr{W}^{1,p}(\mathbb{X}, \mathbb{Y})$ is also closed under weak topology of $\mathscr{W}^{1,p}(\mathbb{X}, \mathbb{R}^N)$.

2.3. Differential forms

Throughout this paper we let $\mathscr{C}^\infty(\wedge^\ell \mathbb{X})$, $0 \leqslant \ell \leqslant n = \dim \mathbb{X}$, denote the space of smooth ℓ-forms on $\mathbb{X}$. Two differential operators on forms will be of particular interest to us. First is the exterior derivative,

$$(2.4) \qquad d : \mathscr{C}^\infty(\wedge^\ell \mathbb{X}) \to \mathscr{C}^\infty(\wedge^{\ell+1} \mathbb{X})$$

Second is the formal adjoint of d, also called Hodge codifferential,

$$(2.5) \qquad d^* = (-1)^{n\ell+1} * d* : \mathscr{C}^\infty(\wedge^{\ell+1}\mathbb{X}) \to \mathscr{C}^\infty(\wedge^\ell \mathbb{X})$$

where $* : \mathscr{C}^\infty(\wedge^\ell \mathbb{X}) \to \mathscr{C}^\infty(\wedge^{n-\ell}\mathbb{X})$ denotes the Hodge star duality operator. Here, we conveniently set $\mathscr{C}^\infty(\wedge^\ell \mathbb{X}) = 0$, whenever $\ell < 0$ or $\ell > n$. Note that $** = (-1)^{\ell(n-\ell)}$ on $\mathscr{C}^\infty(\wedge^\ell \mathbb{X})$. The point-wise scalar product of forms $\alpha, \beta \in \mathscr{C}^\infty(\wedge^\ell \mathbb{X})$ is given by $\langle \alpha, \beta \rangle \, dx = \alpha \wedge *\beta \in \mathscr{C}^\infty(\wedge^n \mathbb{X})$, and hence

$$(2.6) \qquad \int_{\mathbb{X}} \langle \alpha, \beta \rangle \, dx = \int_{\mathbb{X}} \alpha \wedge *\beta$$

The duality between d and d^* is emphasized in the formula of integration by parts

$$(2.7) \qquad \int_{\mathbb{X}} \langle d\varphi, \psi \rangle = \int_{\mathbb{X}} \langle \varphi, d^*\psi \rangle$$

for $\varphi \in \mathscr{C}^\infty(\wedge^\ell \mathbb{X})$ and $\psi \in \mathscr{C}^\infty(\wedge^{\ell+1}\mathbb{X})$. Now a differential form $\varphi \in \mathscr{L}^p(\wedge^\ell \mathbb{X})$ is said to be closed in the sense of distributions if $\int_{\mathbb{X}} \langle \varphi, d^*\psi \rangle = 0$ for every test form $\psi \in \mathscr{C}^\infty(\wedge^{\ell+1}\mathbb{X})$. We write it as $d\varphi = 0$. Similarly, we establish what it means for ψ to be coclosed, and write it as $d^*\psi = 0$. Forms of the type $d\alpha$, with $\alpha \in \mathscr{W}^{1,p}(\wedge^{\ell-1}\mathbb{X})$, are called exact while those of type $d^*\beta$, with $\beta \in \mathscr{W}^{1,p}(\wedge^{\ell+1}\mathbb{X})$, are called coexact. It follows from the identities $d \circ d = 0$ and $d^* \circ d^* = 0$ that the ℓ-forms $d\alpha \in \mathscr{L}^p(\wedge^\ell \mathbb{X})$ and $d^*\beta \in \mathscr{L}^p(\wedge^\ell \mathbb{X})$ are closed and coclosed, respectively. Finally, those forms $h \in \mathscr{L}^p(\wedge^\ell \mathbb{X})$ which are closed and coclosed will be called harmonic fields of degree ℓ. We denote by $\mathcal{H}(\wedge^\ell \mathbb{X})$ the space of all harmonic fields of degree ℓ and regard it as well known that this space is finite dimensional. $\mathcal{H}(\wedge^\ell \mathbb{X})$ consists of $\mathscr{C}^\infty$-smooth forms. Being so, all possible norms on $\mathcal{H}(\wedge^\ell \mathbb{X})$ are equivalent. For instance, we shall benefit from the estimate

$$(2.8) \qquad \|h\|_{\mathscr{L}^\infty(\wedge^\ell \mathbb{X})} \precsim \|h\|_{\mathscr{L}^1(\wedge^\ell \mathbb{X})}$$

and further,

$$(2.9) \qquad \|h\|_{\mathscr{L}^\infty(\wedge^\ell \mathbb{X})} \precsim \left(\int_{\mathbb{X}} |h|^p \right)^{\frac{1}{p}} \qquad \text{for all } p > 0$$

In relation to the imbedding $\mathscr{L}^1_{\text{weak}}(\wedge^\ell \mathbb{X}) \subset \mathscr{L}^p(\wedge^\ell \mathbb{X})$, with $0 < p < 1$, we record the following estimate

$$(2.10) \qquad \|h\|_{\mathscr{L}^\infty(\wedge^\ell \mathbb{X})} \precsim \left(\int_{\mathbb{X}} |h|^p \right)^{\frac{1}{p}} \precsim |\mathbb{X}|^{\frac{1-p}{p}} \sup \left\{ t \int_{|h|>t} dx; \; t > 0 \right\}$$

as is easily verified by Tchebyshev inequality, see (4.6).

2.3.1. Sobolev classes of differential forms. Four spaces of differential forms have a special place in our studies. These spaces are:

- The Sobolev space of closed forms:

$$\mathscr{L}^p(\wedge^\ell \mathbb{X}) \cap \ker d$$

- The Sobolev space of coclosed forms:
$$\mathscr{L}^p(\wedge^\ell \mathbb{X}) \cap \ker d^*$$

- The Sobolev space of exact forms:
$$d\mathscr{W}^{1,p}(\wedge^{\ell-1}\mathbb{X}) = \left\{ d\alpha;\ \alpha \in \mathscr{W}^{1,p}(\wedge^{\ell-1}\mathbb{X}) \right\} \subset \mathscr{L}^p(\wedge^\ell \mathbb{X}) \cap \ker d$$

- The Sobolev space of coexact forms:
$$d^*\mathscr{W}^{1,p}(\wedge^{\ell+1}\mathbb{X}) = \left\{ d^*\beta;\ \beta \in \mathscr{W}^{1,p}(\wedge^{\ell+1}\mathbb{X}) \right\} \subset \mathscr{L}^p(\wedge^\ell \mathbb{X}) \cap \ker d^*$$

It is far from being evident, although it is true, that for $1 < p < \infty$ all four of these classes are complete linear subspaces of $\mathscr{L}^p(\wedge^\ell \mathbb{X})$, see for instance [**35**]. In each of those classes the corresponding subspace of smooth forms is dense.

2.3.2. Hodge decomposition. Decomposition of a differential form $\omega \in \mathscr{L}^p(\wedge^\ell \mathbb{X})$ into exact, coexact and harmonic component will play essential role in our proofs.

THEOREM 2.6. [HODGE DECOMPOSITION] *For $1 < p < \infty$ and $\ell = 0, 1, ..., n$ we have the following direct sum decomposition*

(2.11) $$\mathscr{L}^p(\wedge^\ell \mathbb{X}) = d\mathscr{W}^{1,p}(\wedge^{\ell-1}\mathbb{X}) \oplus d^*\mathscr{W}^{1,p}(\wedge^{\ell+1}\mathbb{X}) \oplus \mathcal{H}(\wedge^\ell \mathbb{X})$$

Accordingly, every $\omega \in \mathscr{L}^p(\wedge^\ell \mathbb{X})$ can be uniquely written as

(2.12) $$\omega = d\alpha + d^*\beta + h$$

where

(2.13) $$\|\alpha\|_{\mathscr{W}^{1,p}(\wedge^{\ell-1}\mathbb{X})} + \|\beta\|_{\mathscr{W}^{1,p}(\wedge^{\ell+1}\mathbb{X})} + \|h\|_{\mathscr{L}^\infty(\wedge^\ell \mathbb{X})} \preccurlyeq \|\omega\|_{\mathscr{L}^p(\wedge^\ell \mathbb{X})}$$

We restrain ourselves to only a few comments about the case when ω is a closed form. It follows from the uniqueness of this decomposition that the coexact component of ω vanishes. That is,

(2.14) $$\omega = d\alpha + h, \quad \text{for } \omega \in \mathscr{L}^p(\wedge^\ell \mathbb{X}) \cap \ker d$$

In fact, this is none other than the $\mathscr{L}^p$-setting of the deRham cohomology:

Every closed form in $\mathscr{L}^p(\wedge^\ell \mathbb{X}) \cap \ker d$ is exact modulo harmonic fields.

In other words, each deRham ℓ-cohomology class of $\mathbb{X}$ is uniquely represented by a harmonic field. In symbols $H^\ell(\mathbb{X}) \cong \mathcal{H}(\wedge^\ell \mathbb{X})$. The harmonic field at (2.14) can be explicitly expressed as $h = \mathbf{T}\omega$, where $\mathbf{T}$ is a Calderón-Zygmund type operator acting on $\mathscr{L}^p(\wedge^\ell \mathbb{X})$. As this operator is weak $(1,1)$-type, a uniform $\mathscr{L}^1_{\text{weak}}$-estimate combined with (2.10) yields

(2.15) $$\|h\|_{\mathscr{L}^\infty(\wedge^\ell \mathbb{X})} \preccurlyeq \sup \left\{ t \int_{|h|>t} dx;\ t > 0 \right\} \preccurlyeq \|\omega\|_{\mathscr{L}^1(\wedge^\ell \mathbb{X})}$$

We reiterate that the implied constant depends on $\mathbb{X}$.

2.3.3. Pullback.

Our applications of the $\mathscr{L}^p$-theory of differential forms concern the pullbacks via a mapping $f : \mathbb{X} \to \mathbb{Y}$ of differential forms in the target space. Let $\alpha \in \mathscr{C}^\infty(\wedge^\ell \mathbb{Y})$ and $f \in \mathscr{W}^{1,p}(\mathbb{X}, \mathbb{Y})$, $p \geqslant \ell \geqslant 1$. The pullback $f^\sharp \alpha$ lies in $\mathscr{L}^{\frac{p}{\ell}}(\wedge^\ell \mathbb{X})$, because of the point-wise inequality

$$|f^\sharp \alpha| \preccurlyeq |Df|^\ell \tag{2.16}$$

We point out that the usual commutation rule

$$f^\sharp \circ d = d \circ f^\sharp \tag{2.17}$$

requires some regularity of f. For instance it holds for $f \in \mathscr{W}^{1,p}(\mathbb{X}, \mathbb{Y})$, provided $p \geqslant \ell + 1$. The exterior derivative in the right hand side is understood in the sense of distributions, while the left hand side is a form in $\mathscr{L}^{\frac{p}{\ell+1}}(\wedge^{\ell+1} \mathbb{X})$.

2.3.4. Partition of unity.

The arguments for our proofs as well as the definition of the Hardy space $\mathscr{H}^1(\mathbb{X})$ will involve a device of regularization. For such purposes partition of unity is needed. To assemble local estimates into global ones we shall make use of partitions of unity with small supports on $\mathbb{X}$.

Given any locally finite covering $\mathfrak{F}$ of $\mathbb{X}$, a smooth partition of unity subordinate to $\mathfrak{F}$ is a collection $\{\varphi_\mathbb{U};\ \mathbb{U} \in \mathfrak{F}\}$ of functions $\varphi_\mathbb{U} \in \mathscr{C}_0^\infty(\mathbb{U})$ such that

- $0 \leqslant \varphi_\mathbb{U}(x) \leqslant 1$
- $\sum_{\mathbb{U} \in \mathfrak{F}} \varphi_\mathbb{U}(x) = 1$ for all $x \in \mathbb{X}$

The existence of partitions of unity is well known. We illustrate the utility of it with an example. Suppose we are given differential forms $\alpha \in \mathscr{L}^p(\wedge^\ell \mathbb{X})$ and $\beta \in \mathscr{W}^{1,p}(\wedge^{\ell-1} \mathbb{X})$. With the aid of a partition of unity we may express α and $d\beta$ as:

$$\alpha = \sum_{\mathbb{U} \in \mathfrak{F}} \alpha_\mathbb{U} \stackrel{\text{def}}{=} \sum_{\mathbb{U} \in \mathfrak{F}} \varphi_\mathbb{U} \, \alpha \tag{2.18}$$

$$d\beta = \sum_{\mathbb{U} \in \mathfrak{F}} d\beta_\mathbb{U} \stackrel{\text{def}}{=} \sum_{\mathbb{U} \in \mathfrak{F}} d(\varphi_\mathbb{U} \, \beta) \tag{2.19}$$

Here each term $d\beta_\mathbb{U}$ is an exact form. In contrast, if α is only closed the terms $\alpha_\mathbb{U}$ are no longer closed forms. In spite of such inconvenience the above decomposition is still useful. The point to it is that the exterior derivative

$$d\alpha_\mathbb{U} = \varphi_\mathbb{U} \, d\alpha + d\varphi_\mathbb{U} \wedge \alpha \tag{2.20}$$

enjoys the same regularity as $d\alpha$.

2.3.5. Cartan forms.

Let $\mathbb{Y}$ be a $\mathscr{C}^\infty$-smooth oriented closed Riemannian manifold of dimension $m \geqslant 2$. Recall that $\mathscr{C}^\infty(\wedge^\ell \mathbb{Y})$, $1 \leqslant \ell \leqslant m$, is a module over the ring $\mathscr{C}^\infty(\mathbb{Y})$. The first thing we wish to discuss here is that $\mathscr{C}^\infty(\wedge^\ell \mathbb{Y})$ is finitely generated by exact differential forms, which we denote by

$$d\Xi_1, d\Xi_2, ..., d\Xi_M \tag{2.21}$$

where $\Xi_i \in \mathscr{C}^\infty(\wedge^{\ell-1} \mathbb{Y})$. This simply means that every $\gamma \in \mathscr{C}^\infty(\wedge^\ell \mathbb{Y})$ can be written as

$$\sum_{i=1}^{M} \lambda_i \, d\Xi_i \quad \text{where } \lambda \in \mathscr{C}^\infty(\mathbb{Y}) \tag{2.22}$$

In general, one cannot guarantee that the generators $d\Xi_1, d\Xi_2, ..., d\Xi_M$ will be linearly independent at each point $y \in \mathbb{Y}$. Therefore, the decomposition at (2.22) need not be unique. Our goal is to select the generators carefully and give an explicit formula for the coefficients $\lambda_i \in \mathscr{C}^\infty(\mathbb{Y})$ in terms of $\gamma \in \mathscr{C}^\infty(\wedge^\ell \mathbb{Y})$.

PROPOSITION 2.7. *There exist differential forms $\Xi_i \in \mathscr{C}^\infty(\wedge^{\ell-1}\mathbb{Y})$ and $\Gamma_i \in \mathscr{C}^\infty(\wedge^\ell \mathbb{Y})$, $i = 1, 2, ..., M$, such that every $\gamma \in \mathscr{C}^\infty(\wedge^\ell \mathbb{Y})$ admits a decomposition*

$$(2.23) \qquad \gamma = \sum_{i=1}^M \lambda_i \, d\Xi_i, \qquad \text{where } \lambda_i = \langle \gamma, \Gamma_i \rangle$$

As before, in (2.6), the symbol $\langle \, , \, \rangle$ stands for the point-wise scalar product of differential forms. Precisely, using the Hodge star operator, $\lambda_i \, dy = \lambda_i \wedge (*\Gamma_i)$. We take particular note of the fact that the functions λ_i depend on γ in a linear fashion.

For the proof, it is useful to begin with a finite atlas of local charts (Ω_k, κ_k) on $\mathbb{Y}$, $k = 1, 2, ..., K$, such that each mapping

$$\kappa_k = (\kappa_k^1, \kappa_k^2, ..., \kappa_k^m) : \Omega_k \to \mathbb{R}^m$$

is a diffeomorphism of an open region $\Omega_k \subset \mathbb{Y}$ onto $\mathbb{R}^m$ and $\bigcup_{k=1}^K \Omega_k = \mathbb{Y}$. Let us state it in this way:

$$d\kappa_k^1 \wedge d\kappa_k^2 \wedge ... \wedge d\kappa_k^m \neq 0 \quad \text{on } \Omega_k$$

Upon obvious modifications (multiply by a suitable bump function) we produce a system of mappings, again denote by κ_k, such that

- Each κ_k is defined on the entire manifold $\mathbb{Y}$ and maps it smoothly into $\mathbb{R}^m$.
- The Jacobian determinant of κ_k, which we define by the rule

$$\mathcal{J}_k(y) \, dy = d\kappa_k^1 \wedge d\kappa_k^2 \wedge ... \wedge d\kappa_k^m$$

satisfies:

$$\mathcal{J}_k(y) \geqslant 1 \qquad \text{for } y \in \Omega_k$$

In this extension of κ_k the new open sets Ω_k are actually slightly smaller than the original ones, though they still cover the manifold $\mathbb{Y}$. To each $k = 1, 2, ..., K$ and ℓ-tuple I; $1 \leqslant i_1 < i_2 < ... < i_\ell \leqslant m$, there corresponds a differential form

$$(2.24) \qquad \Xi_k^I = \kappa_k^{i_1} \, d\kappa_k^{i_2} \wedge ... \wedge d\kappa_k^{i_\ell} \in \mathscr{C}^\infty(\wedge^{\ell-1}\mathbb{Y})$$

Now, the exact forms that will generate $\mathscr{C}^\infty(\wedge \mathbb{Y})$, can be defined as

$$(2.25) \qquad d\Xi_k^I = d\kappa_k^{i_1} \wedge d\kappa_k^{i_2} \wedge ... \wedge d\kappa_k^{i_\ell} \in \mathscr{C}^\infty(\wedge^\ell \mathbb{Y})$$

They generate the module $\mathscr{C}^\infty(\wedge \mathbb{Y})$ over the ring $\mathscr{C}^\infty(\mathbb{Y})$. To see this we fix a partition of unity $\{\varphi_k\}_{k=1,2,...,K}$ subordinate to the cover $\{\Omega_k\}_{k=1,2,...,K}$. Let us introduce differential forms $\Gamma_k^I \in \mathscr{C}_\circ^\infty(\wedge^\ell \Omega_k)$, by the rule

$$(2.26) \qquad *\Gamma_k^I = \frac{\varphi_k \, d\Xi_k^{I'}}{\mathcal{J}_k}, \qquad \text{where} \qquad \begin{array}{c} I: 1 \leqslant i_1 < i_2 < ... < i_\ell \leqslant m \\ k = 1, 2, ..., K \end{array}$$

The superscript I' stands for an ordered complementary $(m - \ell)$-tuple; $I' = (i_1', i_2', ..., i_{m-\ell}')$ is ordered in such a way that $(i_1, ..., i_\ell, i_1', ..., i_{m-\ell}')$ constitutes an even permutation of $(1, 2, ..., m)$. In this way $d\Xi_k^{I'} = d\kappa_k^{i_1'} \wedge ... \wedge d\kappa_k^{i_{m-\ell}'}$ is a smooth $(m-\ell)$-form on $\mathbb{Y}$. The Hodge star operator $* : \mathscr{C}^\infty(\wedge^{m-\ell}\mathbb{Y}) \to \mathscr{C}^\infty(\wedge^\ell \mathbb{Y})$ converts it onto an ℓ-form, thus $\Gamma_k^I \in \mathscr{C}_\circ^\infty(\wedge^\ell \Omega_k)$. To prove the decomposition formula at (2.23) we recall that $\varphi_k \gamma$, being a form in $\mathscr{C}^\infty(\wedge^\ell \Omega_k)$, can be uniquely written as

$$\varphi_k \gamma = \sum_{1 \leqslant i_1 < ... < i_\ell \leqslant m} \alpha_k^{i_1...i_\ell} \, d\kappa_k^{i_1} \wedge d\kappa_k^{i_2} \wedge ... \wedge d\kappa_k^{i_\ell} \quad \text{where } \alpha_k^{i_1...i_\ell} \in \mathscr{C}^\infty(\Omega_k)$$

To compute the coefficients $\alpha_k^{i_1,...,i_\ell}$ we take the wedge product of both sides with the form,

$$d\Xi_k^{I'} = d\kappa_k^{i'_1} \wedge d\kappa_k^{i'_2} \wedge ... \wedge d\kappa_k^{i'_{m-\ell}} \in \mathscr{C}^\infty(\wedge^\ell \mathbb{Y})$$

The exterior multiplication annihilates all terms except the one corresponding to $I; 1 \leqslant i_1 < ... < i_\ell \leqslant m$

$$\varphi_k \gamma \wedge d\Xi_k^{I'} = \alpha_k^{i_1...i_\ell}(y) \, \mathcal{J}_k(y) \, dy$$

Applying Hodge star operator this equation gives us the coefficients

(2.27) $$\alpha_k^{i_1...i_\ell} = * \left(\gamma \wedge \frac{\varphi_k \, d\Xi_k^{I'}}{\mathcal{J}_k} \right) = \langle \gamma, \Gamma_k^I \rangle$$

Hence

(2.28) $$\varphi_k \gamma = \sum_I \langle \gamma, \Gamma_k^I \rangle \, d\Xi_k^I, \quad k = 1, 2, ..., K$$

Finally, summing up, we arrive at the desired decomposition

(2.29) $$\gamma = \sum_{k=1}^K \sum_I \langle \gamma, \Gamma_k^I \rangle \, d\Xi_k^I$$

However, to be consistent with the assertion of Proposition 2.8 we must rename the indices. Precisely, we must abbreviate the multi-index $\begin{smallmatrix} I \\ k \end{smallmatrix}$ to a single letter $i = 1, 2, ..., M$, where $M = K \binom{m}{\ell}$.

Uniform bounds of the functions λ_i in terms of γ follow directly from the formula (2.23). Let us record the following one

(2.30) $$\| \lambda_i \|_{\mathscr{C}^1(\mathbb{Y})} \precsim \| \gamma \|_{\mathscr{C}^1(\mathbb{Y})}$$

The decomposition, as constructed above, is at intrinsic interest in regard to the following representation of the exact forms on $\mathbb{Y}$.

PROPOSITION 2.8. *Every exact n-form $\omega \in \mathscr{C}^\infty(\wedge^n \mathbb{Y})$, $2 \leqslant n \leqslant \dim \mathbb{Y}$, can be written as*

(2.31) $$\omega = \sum_{i=1}^M d\lambda_i \wedge d\Xi_i$$

where λ_i are smooth functions on $\mathbb{Y}$.

PROOF. We write $\omega = d\gamma$, with some $\gamma \in \mathscr{C}^\infty(\wedge^{n-1} \mathbb{Y})$. Then, with the aid of Proposition 2.7, we decompose γ as

$$\gamma = \sum_{i=1}^K \lambda_i \, d\Xi_i$$

Finally, we differentiate to obtain

$$d\gamma = \sum_{i=1}^K d\lambda_i \wedge d\Xi_i$$

as desired.

Concerning estimates of λ_i in terms of ω, let us emphasize that our decomposition depends on the choice of γ. For this reason it is desirable to introduce the following norm

$$[\![\omega]\!]_\infty = \inf \left\{ \|\gamma\|_{\mathscr{C}^1(\wedge^{n-1}\mathbb{Y})}; \ d\gamma = \omega \right\} \qquad (2.32)$$

With this definition at hand we can achieve the following estimate

$$\|\lambda_i\|_{\mathscr{C}^1(\mathbb{Y})} \precsim [\![\omega]\!]_\infty \qquad (2.33)$$

Proposition 2.8 gives rise to a class of the so called Cartan forms, named after H. Cartan who studied similar differential expressions.

DEFINITION 2.9. [CARTAN FORMS] An n-form $\omega \in \mathscr{C}^\infty(\wedge^n \mathbb{Y})$, $n \leqslant m = \dim \mathbb{Y}$, is said to be a Cartan form if it can be decomposed as

$$\omega = \sum_{i=1}^{M} \alpha_i \wedge \beta_i \qquad (2.34)$$

where $\alpha_i \in \mathscr{C}^\infty(\wedge^{\ell_i} \mathbb{Y}) \cap \ker d$ and $\beta_i \in \mathscr{C}^\infty(\wedge^{k_i} \mathbb{Y}) \cap \ker d$. Here we assume that $k_i, \ell_i \geqslant 1$ and $k_i + \ell_i = n$.

Thus the exact forms in $\mathscr{C}^\infty(\wedge^n \mathbb{Y})$, with $2 \leqslant n \leqslant \dim \mathbb{Y}$, are Cartan forms. We take up this topic here by assuming that $\dim \mathbb{Y} = \dim \mathbb{X} = n$.

COROLLARY 2.10. *Every n-form $\omega \in \mathscr{C}^\infty(\wedge^n \mathbb{Y})$, $\dim \mathbb{Y} = n$, whose integral over $\mathbb{Y}$ vanishes is exact. Consequently, ω is a Cartan form of the type*

$$\omega = \sum_{i=1}^{M} d\lambda_i \wedge d\Xi_i \qquad (2.35)$$

COROLLARY 2.11. *If the manifold $\mathbb{Y}$ of dimension n admits at least one Cartan n-form with non-vanishing integral, then all n-forms on $\mathbb{Y}$ are Cartan forms.*

COROLLARY 2.12. [DECOMPOSITION OF n-FORMS] *Every $\omega \in \mathscr{C}^\infty(\wedge^n \mathbb{Y})$, $\dim \mathbb{Y} = n$, can be written as*

$$\omega = \left(\fint_{\mathbb{Y}} \omega \right) dy + \sum_{i=1}^{M} d\lambda_i \wedge d\Xi_i \qquad (2.36)$$

Next we bring on stage the manifolds which are cohomologically simple. Recall that $\mathbb{Y}$ is a rational homology sphere if all its cohomology groups $H^\ell(\mathbb{Y})$, with $1 \leqslant \ell < n$, vanish. In this case Cartan n-forms

$$\omega = \sum_{i=1}^{M} \alpha_i \wedge \beta_i \qquad (2.37)$$

are necessarily exact and as such have vanishing integral mean over $\mathbb{Y}$. Indeed, every closed form α_i is exact so is each wedge product $\alpha_i \wedge \beta_i$, $i = 1, ..., M$. In other words, the condition $H^\ell(\mathbb{Y}) \neq 0$, for some $1 \leqslant \ell < n$, is necessary in order to find a Cartan form on $\mathbb{Y}$ with non-vanishing integral. Our next result shows that this condition is also sufficient.

Suppose $H^\ell(\mathbb{Y}) \neq 0$ for some $1 \leqslant \ell < n$. Hodge-deRham theory tells us that there exists a nonzero harmonic field $h \in \mathcal{H}(\wedge^\ell \mathbb{Y})$. Consider the n-form

$$\omega = h \wedge *h = |h|^2 \, dy \qquad (2.38)$$

where $*h \in \mathscr{C}^\infty(\wedge^{n-\ell}\mathbb{Y})$ is Hodge dual to h. By the definition of $\mathcal{H}(\wedge^\ell \mathbb{Y})$ the form h is both closed and coclosed. We then see that $*h$ is also closed. Thus ω is a Cartan form. That ω has non-vanishing integral follows from the formula

$$\int_\mathbb{Y} \omega = \int_\mathbb{Y} h \wedge *h = \int_\mathbb{Y} |h|^2 \, dy \neq 0 \tag{2.39}$$

We end this section by combining these later observations with Corollary 2.11.

PROPOSITION 2.13. *Let* $\dim \mathbb{Y} = n$. *Then every* $\omega \in \mathscr{C}^\infty(\wedge^n \mathbb{Y})$ *is a Cartan form if and only if* $H^\ell(\mathbb{Y}) \neq 0$, *for some* $1 \leqslant \ell < n$; *that is, if* $\mathbb{Y}$ *is not rational homology sphere.*

2.4. Mollifiers and smoothing operator

For the duration of this paper we fix a nonnegative function $\boldsymbol{\Phi} \in \mathscr{C}_0^\infty(\mathbb{R}^n)$ supported in the closed unit ball and having integral 1. For example

$$\boldsymbol{\Phi}(x) = C_n \begin{cases} \exp \frac{1}{|x|^2 - 1} & \text{if } |x| < 1 \\ 0 & \text{if } |x| \geqslant 1 \end{cases} \tag{2.40}$$

where C_n is a constant. The one parameter family

$$\boldsymbol{\Phi}_t(x) = \frac{1}{t^n} \boldsymbol{\Phi}\left(\frac{x}{t}\right), \quad t > 0 \tag{2.41}$$

defines an approximation of the Dirac mass at the origin; $\boldsymbol{\Phi}_t$ are called mollifiers. Given $u \in \mathscr{L}^1_{\text{loc}}(\mathbb{R}^n)$, the mollification of u is the family of functions $u_t \in \mathscr{C}^\infty(\mathbb{R}^n)$, $t > 0$, defined by the convolution formula

$$u_t(x) = \int_{\mathbb{R}^n} \boldsymbol{\Phi}_t(x-z) u(z) \, dz \tag{2.42}$$

Various bounds for a function $u \in \mathscr{W}^{1,p}_{\text{loc}}(\mathbb{R}^n)$, $1 \leqslant p \leqslant \infty$, imply the same bounds for u_t. Basic properties of the mollification are listed below:
   (i) $\lim_{t \to 0} u_t(x) = u(x)$ for almost every $x \in \mathbb{R}^n$
   (ii) If u is continuous then u_t converges to u uniformly on compact subsets
   (iii) Mollification preserves the $\mathscr{L}^p$-bounds;

$$\|u_t\|_{\mathscr{L}^p(\mathbb{R}^n)} \precsim \|u\|_{\mathscr{L}^p(\mathbb{R}^n)}, \quad 1 \leqslant p \leqslant \infty$$

$$\|du_t\|_{\mathscr{L}^p(\mathbb{R}^n)} \precsim \|du\|_{\mathscr{L}^p(\mathbb{R}^n)}, \quad 1 \leqslant p \leqslant \infty$$

   (iv) For $1 \leqslant p < \infty$ and $u \in \mathscr{W}^{1,p}(\mathbb{R}^n)$, we have

$$\lim_{t \to 0} \|u_t - u\|_{\mathscr{L}^p(\mathbb{R}^n)} = 0$$

$$\lim_{t \to 0} \|u_t - u\|_{\mathscr{W}^{1,p}(\mathbb{R}^n)} = 0$$

The implied constants in (iii) are actually equal to 1, but not on manifolds latter on. It is well known that $\lim_{t \to 0} u_t(x) = u(x)$ at every Lebesgue point of u, regardless of the generating mollifier $\boldsymbol{\Phi}$. For this property one could take $\boldsymbol{\Phi}$ to be the normalized characteristic function of the unit ball:

$$\boldsymbol{\Phi}(x) = \frac{\chi_\mathbb{B}(x)}{|\mathbb{B}|} \tag{2.43}$$

But this choice of $\mathbf{\Phi}$ does not work for the definition of the Hardy space. Nevertheless, for $\mathbf{\Phi}$ defined at (2.43), the mollifications lead us to the familiar $\mathscr{L}^1$-averages over the balls $\mathbb{B}(x,r)$:

$$u_r(x) = \fint_{\mathbb{B}(x,r)} u(z)\, dz \tag{2.44}$$

Next we invoke the reference atlas $\mathcal{A}$ on $\mathbb{X}$, which we have already fixed in Section 2.1.1. Now we also fix a partition of unity $\{\varphi_\Omega\}_{\Omega \in \mathcal{A}}$ subordinate to $\mathcal{A}$. Thus, to every Ω there corresponds a coordinate mapping $\kappa : \Omega \xrightarrow{onto} \mathbb{R}^n$. The mollification of a function $f \in \mathscr{L}^1(\mathbb{X})$ can now be defined by the rule

$$f_t(x) = \sum_{\Omega \in \mathcal{A}} \varphi_\Omega(x) \int_\Omega \mathbf{\Phi}_t(\kappa(x) - \kappa(z)) \mathcal{J}(z, \kappa) f(z)\, dz \tag{2.45}$$

Each integral term is a smooth function on Ω, equal to $(f \circ \kappa^{-1})_t \circ \kappa \in \mathscr{C}^\infty(\Omega)$. We shall write this formula in a compact form as

$$f_t(x) = \int_\mathbb{X} K_t(x, z) f(z)\, dz \tag{2.46}$$

where $K_t : \mathbb{X} \times \mathbb{X} \to \mathbb{R}_+$ is $\mathscr{C}^\infty$-smooth for all $t > 0$.

Clearly, all basic properties of the mollification procedure, listed in (i-iv), remain valid on a manifold $\mathbb{X}$. But we must restrict ourselves to sufficiently small parameters t, say

$$0 < t \leqslant t_\mathbb{X} \tag{2.47}$$

The upper bound $t_\mathbb{X}$ depends not only on $\mathbb{X}$, but also on the atlas $\mathcal{A}$ and the partition of unity. As those entities are fixed once and for all, $t_\mathbb{X}$ is also fixed for the duration of this text. The reader may wish to note that the proof of (iii) relies on Lemma 2.3 with $\mathbb{E} = \mathbb{X}$.

Mollification procedure usually expands the support of f. For example, if $\operatorname{supp} f \subset \mathbb{U}$, then $\operatorname{supp} f_t \subset \mathbb{U}_{t'}$, where

$$\mathbb{U}_{t'} = \{x \in \mathbb{X}\,;\ \operatorname{dist}(x, \mathbb{U}) < t'\} \quad \text{and} \quad t \preccurlyeq t' \preccurlyeq t$$

Here too, the implied constants for the inequalities $t \preccurlyeq t' \preccurlyeq t$ depend only on $\mathbb{X}$. This can be seen from the equation

(v) $K_t(x, z) = 0$, whenever $\operatorname{dist}(x, z) \succcurlyeq t$

Moreover, if $f : \mathbb{X} \to \mathbb{R}^N$ is constant then $f_t : \mathbb{X} \to \mathbb{R}^N$ is also constant for all $t > 0$, which is immediate from the identity

(vi) $\int_\mathbb{X} K_t(x, z)\, dz \equiv 1$, for all $t > 0$

Finally, combining (v) and (vi) we obtain

(vii) $\operatorname*{osc}_\mathbb{U} f_t \preccurlyeq \operatorname*{ess\,osc}_{\mathbb{U}'} f \quad t \preccurlyeq t' \preccurlyeq t$

Perhaps, the definition of the essential oscillation of a measurable function $f : \mathbb{V} \to \mathbb{R}^N$ is in order. The symbol $\operatorname*{ess\,osc}_\mathbb{V} f$ stands for the infimum of all $\delta > 0$ such that the set $\{(x_1, x_2) \in \mathbb{V} \times \mathbb{V};\ |f(x_1) - f(x_2)| > \delta\}$ has measure zero in $\mathbb{V} \times \mathbb{V}$.

2.5. Maximal operators

The well-developed study of maximal functions has an analogue for Riemannian manifolds. Here we shall frame the definitions and basic properties in this setting, some pending a discussion in subsequent sections.

2.5.1. The Hardy-Littlewood maximal operator.
Given $1 \leqslant p < \infty$ and $h \in \mathscr{L}^p(\mathbb{X}, \mathbf{V})$, where $\mathbf{V}$ is a finite dimensional normed space, we define

$$(2.48) \qquad \mathbf{M}_p h(x) = \sup\left\{ \left(\fint_{\mathbb{B}} |h(x)|^p \, dx\right)^{\frac{1}{p}}; \; x \in \mathbb{B} \subset \mathbb{X} \right\}$$

The supremum runs over all metric balls $\mathbb{B} \subset \mathbb{X}$ containing a given point x. Since the entire manifold $\mathbb{X}$ is also a ball we see that for every $x \in \mathbb{X}$

$$(2.49) \qquad \mathbf{M}_p h(x) \geqslant \left(\fint_{\mathbb{X}} |h|^p\right)^{\frac{1}{p}}$$

Lebesgue Differentiation Theorem tells us that

$$(2.50) \qquad |h(x)| \leqslant \mathbf{M}_p h(x) \quad \text{for a.e. } x \in \mathbb{X}$$

Also note, by using Hölder's inequality, that the function $p \to \mathbf{M}_p h(x)$ is increasing. For notational convenience we omit the subscript p if it equals 1. Thus

$$(2.51) \qquad \mathbf{M}_p h = (\mathbf{M} |h|^p)^{\frac{1}{p}}$$

PROPOSITION 2.14. [WEAK TYPE ESTIMATE] *For every $h \in \mathscr{L}^p(\mathbb{X}, \mathbf{V})$ the maximal function $\mathbf{M}_p h$ belongs to the Marcinkiewicz class $\mathscr{L}^p_{weak}(\mathbb{X})$ and we have*

$$(2.52) \qquad \int_{\mathbf{M}_p h > 2t} dx \; \preccurlyeq \; \frac{1}{t^p} \int_{|h| > t} |h(x)|^p \, dx$$

for all $t > 0$.

We will not prove this proposition here. The arguments establishing (2.52) are very similar to those used in the Euclidean setting. Let us point out that the main tool is Vitali type covering lemma, which is true in any separable metric space. But we are not involved in such generality. The interested reader may try to consult [**12**, 2.8.4-2.8.6].

As a consequence of the weak-type estimate at (2.52), and of sublinearity of $\mathbf{M}_s$, we obtain

COROLLARY 2.15. *Let $\{h_j\}$ converge to h in $\mathscr{L}^s(\mathbb{X}, \mathbf{V})$, $s \geq 1$, then $\{\mathbf{M}_s h_j\}$ contains a subsequence converging to $\mathbf{M}_s h$ almost everywhere.*

Of course, $\mathbf{M}_p$ is a sublinear bounded operator in $\mathscr{L}^\infty(\mathbb{X}, \mathbf{V})$. By interpolation, we infer strong type estimates.

PROPOSITION 2.16. *The maximal operator $\mathbf{M}_p : \mathscr{L}^s(\mathbb{X}, \mathbf{V}) \to \mathscr{L}^s(\mathbb{X})$ is bounded for all $p < s \leqslant \infty$. Precisely, we have*

$$(2.53) \qquad \|\mathbf{M}_p h\|_s \; \preccurlyeq \; \|h\|_s$$

Reviewing the maximal function $\mathbf{M}f$ in relation to the mollifiers f_t we first notice that

$$(2.54) \qquad |f_t(x)| \; \preccurlyeq \; \fint_{\mathbb{B}(x,r)} |f(z)| \, dz$$

where $t \leqslant r \preccurlyeq t$. In particular,

$$(2.55) \qquad |f_t(x)| \; \preccurlyeq \; \mathbf{M}f(x) \quad \text{for every } x \in \mathbb{X}$$

As for the differential Df_t, we have the following estimate

$$(2.56) \qquad |Df_t(x)| \; \preccurlyeq \; \fint_{\mathbb{B}(x,r)} |f(z)| \, dz + \fint_{\mathbb{B}(x,r)} |Df(z)| \, dz$$

This also applies to $f(x) - C$, where C can be any constant. It has to be noted that $(f-C)_t = f_t - C$, hence

$$(2.57) \qquad |Df_t(x)| \preccurlyeq \fint_{\mathbb{B}(x,r)} |f(z) - C|\, dz + \fint_{\mathbb{B}(x,r)} |Df(z)|\, dz$$

Poincaré inequality yields a desired estimate of the first integral in terms of Df.

$$\fint_{\mathbb{B}(x,r)} |f(z) - C|\, dz \preccurlyeq r \fint_{\mathbb{B}(x,r)} |Df(z)|\, dz \preccurlyeq \fint_{\mathbb{B}(x,r)} |Df(z)|\, dz$$

Finally, as the integral averages do not exceed the maximal function, we conclude with the inequality

$$(2.58) \qquad |Df_t(x)| \preccurlyeq \mathbf{M}(Df)(x) \quad \text{for all } x \in \mathbb{X}$$

2.5.2. The Fefferman-Stein operator and the Hardy space. The one parameter family of $\mathscr{C}^\infty$-smooth functions $K_t : \mathbb{X} \times \mathbb{X} \to \mathbb{R}_+$, $0 < t \leqslant t_\mathbb{X}$, introduced in Section 2.4, will be employed to define another maximal operator on $\mathbb{X}$. For an n-form $\omega \in \mathscr{L}^1(\wedge^n \mathbb{X})$ we define a function $\omega_t \in \mathscr{C}^\infty(\mathbb{X})$ by the rule

$$(2.59) \qquad \omega_t(x) = \int_\mathbb{X} K_t(x, \cdot)\, \omega, \quad 0 < t \leqslant t_\mathbb{X}$$

If $\omega = h(z)\, dz$, where h is an integrable function and dz is the Riemannian volume element on $\mathbb{X}$, we also write

$$(2.60) \qquad h_t(x) = \int_\mathbb{X} K_t(x, z)\, h(z)\, dz$$

Then $\lim_{t \to 0} h_t(x) = h(x)$ at the Lebesgue points of h. This gives a way to the concept of the maximal operator; replace $\lim_{t \to 0}$ by $\sup_{t > 0}$. Recall that the Hardy-Littlewood maximal function of $h \in \mathscr{L}^1(\mathbb{X})$ is defined by

$$(2.61) \qquad (\mathbf{M}h)(x) = \sup \left\{ \fint_\mathbb{B} |h(z)|\, dz;\ x \in B \subset \mathbb{X} \right\} \succcurlyeq \sup_{0 < t \leqslant t_\mathbb{X}} |h|_t(x)$$

where the supremum runs over all metric balls containing the given point $x \in \mathbb{X}$. More sensitive on various cancellations is the Fefferman-Stein maximal function

$$(2.62) \qquad (\mathcal{M}h)(x) = \sup_{0 < t \leqslant t_\mathbb{X}} |h_t(x)|, \quad h \in \mathscr{L}^1(\wedge^n \mathbb{X})$$

Let us emphasize explicitly that here we first mollify h and then take the absolute value of it. Clearly, we have

$$(2.63) \qquad \mathcal{M}h(x) \preccurlyeq \mathbf{M}h(x)$$

but not conversely. As a note of additional interest, the maximal operator $\mathcal{M}$ can be defined on Schwartz distributions due to the smoothness of the generating function Φ, see [**53**].

DEFINITION 2.17. [HARDY SPACE] An n-form $h \in \mathscr{L}^1(\wedge^n \mathbb{X})$ is said to be in the Hardy space $\mathscr{H}^1(\wedge^n \mathbb{X})$ if $\mathcal{M}h \in \mathscr{L}^1(\mathbb{X})$.

We see that $\mathscr{H}^1(\wedge^n \mathbb{X})$ is a Banach space with respect to the norm

$$(2.64) \qquad \|h\|_{\mathscr{H}^1(\mathbb{X})} = \int_\mathbb{X} \mathcal{M}h$$

We refer to [**54**] for yet another approach to $\mathscr{H}^1$-spaces on manifolds. Now, we recall very briefly the Zygmund space $\mathscr{L}\log\mathscr{L}(\mathbb{X})$. It consists of functions $h : \mathbb{X} \to \mathbb{R}$ such that

$$(2.65) \qquad \|h\|_{\mathscr{L}\log\mathscr{L}} = \int_{\mathbb{X}} |h(x)| \log\left(e + \frac{|h(x)|}{\int_{\mathbb{X}} |h|}\right) dx < \infty$$

see also Section 4.3. It is known that $\mathscr{L}\log\mathscr{L}(\mathbb{X}) \subset \mathscr{H}^1(\mathbb{X})$. Indeed, for $h \in \mathscr{L}\log\mathscr{L}(\mathbb{X})$, we have the point-wise inequality $|\mathcal{M}h| \leqslant |\mathbf{M}h| \in \mathscr{L}^1(\mathbb{X})$, by Stein's Theorem [**52**]. Conversely, any non-negative function in $\mathscr{H}^1(\mathbb{X})$ lies in $\mathscr{L}\log\mathscr{L}(\mathbb{X})$, see [**53**].

CHAPTER 3

Examples

In this section we go through some well known and some new examples which provide a view on weakly differentiable mapping. In this vein the following modification of the example by R. Schoen and K. Uhlenbeck [49] proves to be most desirable.

3.1. The longitude projection

Consider the n-sphere $\mathbb{S}^n$ in the Euclidean space $\mathbb{R}^{n+1} = \mathbb{R}^n \times \mathbb{R}$. We write the point $x \in \mathbb{S}^n$ as $x = (\mathbf{z}, x_{n+1})$, where $\mathbf{z} \in \mathbb{R}^n$ and $x_{n+1} \in \mathbb{R}$ are coupled by the equation $|\mathbf{z}|^2 + |x_{n+1}|^2 = 1$. The projection along the longitude lines of $\mathbb{S}^n$ onto its equatorial sphere $\mathbb{S}^{n-1} = \{y \in \mathbb{R}^n;\ |y| = 1\}$ is defined by the rule

$$(3.1) \qquad f : \mathbb{S}^n \setminus \{\mathfrak{n}, \mathfrak{s}\} \to \mathbb{S}^{n-1}, \quad f(\mathbf{z}, x_{n+1}) = \frac{\mathbf{z}}{|\mathbf{z}|}$$

Thus f is not defined at the north pole $\mathfrak{n} = (\mathbf{0}, 1)$ and the south pole $\mathfrak{s} = (\mathbf{0}, -1)$. Elementary computation shows that

$$(3.2) \qquad |Df(x)| = \frac{1}{|\mathbf{z}|}$$

Therefore, $f \in \mathscr{W}^{1,p}(\mathbb{S}^n, \mathbb{S}^{n-1})$ for all $1 \leqslant p < n$ but not for $p = n$. Actually, its differential lies in the Marcinkiewicz space $\mathscr{L}^n_{\text{weak}}(\mathbb{X})$. Precisely, we have

$$(3.3) \qquad \frac{1}{t^n} \preccurlyeq \int_{|Df|>t} dx \preccurlyeq \frac{1}{t^n}, \quad \text{for } t \geqslant 1$$

Arguing by analytic methods of topological degree we find that f cannot be approximated by smooth mappings

$$f_j = (f_j^1, f_j^2, ..., f_j^n) : \mathbb{S}^n \to \mathbb{S}^{n-1}$$

in the metric of $\mathscr{W}^{1,n-1}(\mathbb{S}^n, \mathbb{S}^{n-1})$. Indeed, looking for a contradiction we examine the wedge products

$$df_j^1 \wedge ... \wedge df_j^n \in \mathscr{C}^\infty(\wedge^n \mathbb{S}^n)$$

The 1-forms $df_j^1, df_j^2, ..., df_j^n$ are linearly dependent (at each point) because of the relation $|f_j^1|^2 + ... + |f_j^n|^2 = 1$, which yields that $f_j^1 df_j^1 + ... + f_j^n df_j^n \equiv 0$. Hence, the wedge products $df_j^1 \wedge ... \wedge df_j^n$ are identically equal to zero. Now, for every $\varphi \in \mathscr{C}^\infty(\mathbb{S}^n)$ the integration by parts yields

$$0 = \int_{\mathbb{S}^n} \varphi\, df_j^1 \wedge ... \wedge df_j^n = -\int_{\mathbb{S}^n} f_j^1\, d\varphi \wedge df_j^2 \wedge ... \wedge df_j^n$$

The interested reader may recognize that the right hand side defines the so-called distributional wedge product [27], [32]. These latter integrands possess sufficient degree of integrability to pass the limit under the integral sign as $j \to \infty$. Elementary computation then shows that

$$|\mathbb{B}^n| \left[\varphi(\mathfrak{n}) - \varphi(\mathfrak{s}) \right] = \int_{\mathbb{S}^n} f^1 d\varphi \wedge df^2 \ldots \wedge df^n = \lim_{j \to \infty} \int_{\mathbb{S}^n} f_j^1 d\varphi \wedge df_j^2 \ldots \wedge df_j^n = 0,$$

which gives a clear contradiction if we choose $\varphi(\mathfrak{n}) \neq \varphi(\mathfrak{s})$.

In this example f fails to be uniformly continuous near the poles. Furthermore, the oscillation of f on any $(n-1)$-surface surrounding one of those poles stays away from zero, no matter how close to the pole the surface is. And this is exactly what we shall try to avoid. But the precise results must wait until the relevant concepts will be introduced.

3.2. Spherical coordinates

Let $x = (x_1, x_2, ..., x_{n+1})$ be a point in the n-sphere $\mathbb{S}^n \subset \mathbb{R}^{n+1}$, $x_1^2 + ... + x_{n+1}^2 = 1$. Spherical coordinates $(\mathbf{z}, \theta)$ on $\mathbb{S}^n$ can be introduced by setting

(3.4) $$(x_1, x_2, ..., x_{n+1}) = (\mathbf{z} \sin \theta, \cos \theta)$$

where $\mathbf{z}$ lies in the equatorial sphere $\mathbb{S}^{n-1} \subset \mathbb{R}^n$, and $0 \leqslant \theta \leqslant \pi$ is the latitude angle, $x_{n+1} = \cos \theta$. Singularities occur at the north pole $\mathfrak{n} = (0, ..., 0, 1)$, $\theta = 0$, and at the south pole $\mathfrak{s} = (0, ..., 0, -1)$, $\theta = \pi$. At those poles we cannot determine the equatorial coordinate $\mathbf{z} \in \mathbb{S}^{n-1}$. The usual volume element on $\mathbb{S}^n$ takes the following form in spherical coordinates

(3.5) $$dx = |\sin \theta|^{n-1} d\theta \, d\mathbf{z}$$

where $d\mathbf{z}$ stands for the standard volume element on $\mathbb{S}^{n-1}$. Thus, in particular

(3.6) $$\omega_n = |\mathbb{S}^n| = \int_{\mathbb{S}^n} dx = \omega_{n-1} \int_0^\pi \sin^{n-1} \theta \, d\theta$$

For $0 \leqslant \alpha < \beta \leqslant \pi$ we shall consider the spherical slice

(3.7) $$\mathbb{S}_\alpha^\beta = \{ (\mathbf{z} \cos \theta, \sin \theta); \ \mathbf{z} \in \mathbb{S}^{n-1} \text{ and } \alpha \leqslant \theta \leqslant \beta \}$$

Its n-dimensional volume can be estimated by

(3.8) $$|\mathbb{S}_\alpha^\beta| = \omega_{n-1} \int_\alpha^\beta \sin^{n-1} \theta \, d\theta \leqslant \frac{\omega_{n-1}}{n} (\beta^n - \alpha^n) = |\mathbb{B}^n|(\beta^n - \alpha^n)$$

3.3. Winding around the longitude circles

Given a spherical slice $\mathbb{S}_\alpha^\beta$, $0 \leqslant \alpha < \beta \leqslant \pi$, let $\gamma : [\alpha, \beta] \to [0, \infty)$ be an increasing function. A mapping $f : \mathbb{S}_\alpha^\beta \to \mathbb{S}^n$, defined using spherical coordinates by the rule

(3.9) $$f(\mathbf{z} \sin \theta, \cos \theta) = (\mathbf{z} \sin \gamma(\theta), \cos \gamma(\theta)) \quad \text{for } \alpha \leqslant \theta \leqslant \beta,$$

is called winding map. Note that $(\mathbf{z} \sin \theta, \cos \theta)$ and $(\mathbf{z} \sin \gamma(\theta), \cos \gamma(\theta))$ lay on the same longitude circle.

We calculate the Jacobian determinant $\mathcal{J}(x, f)$ at the points where the derivative $\gamma'(\theta)$ exists. Observe that the linear tangent map $Df(x) : T_x \mathbb{S}^n \to T_{f(x)} \mathbb{S}^n$ is stretching in the longitude direction by $\gamma'(\theta)$ and in all the equatorial directions by the factor $\left| \frac{\sin \gamma(\theta)}{\sin \theta} \right|$. Hence the Jacobian determinant of f at $x = (\mathbf{z} \cos \theta, \sin \theta)$ is the product of those stretching factors.

(3.10) $$\mathcal{J}(x, f) = \gamma'(\theta) \left| \frac{\sin \gamma(\theta)}{\sin \theta} \right|^{n-1} \geqslant 0$$

The operator norm of the linear map $Df(x)$ is precisely equal to:

$$(3.11) \qquad |Df(x)| = \max\left\{\gamma'(\theta), \left|\frac{\sin\gamma(\theta)}{\sin\theta}\right|\right\}, \qquad x = (\mathbf{z}\sin\theta, \cos\theta)$$

3.4. A mapping of infinite degree

Consider a sequence of latitude angles

$$(3.12) \qquad \theta_0 = \pi > \frac{\pi}{2} \geqslant 2\theta_1 > \theta_1 \geqslant 2\theta_2 > \theta_2 \geqslant \ldots 2\theta_k > \theta_k \geqslant 2\theta_{k+1} > \theta_{k+1}\ldots > 0$$

Additional conditions on these angles will be imposed later on. The n-sphere $\mathbb{S}^n$ is divided into countable number of spherical slices

$$(3.13) \qquad \mathbb{S}^n = \bigcup_{k=0}^{\infty} \mathbb{S}_{2\theta_{k+1}}^{\theta_k} \cup \bigcup_{k=1}^{\infty} \mathbb{S}_{\theta_k}^{2\theta_k}$$

We now construct an infinite covering $f : \mathbb{S}^n \to \mathbb{S}^n$ by the rule

- $f : \mathbb{S}_{2\theta_{k+1}}^{\theta_k} \to \mathbb{S}^n$ is the identity for $k = 0, 1, \ldots$
- $f : \mathbb{S}_{\theta_k}^{2\theta_k} \to \mathbb{S}^n$ is the latitude winding for $k = 1, 2, \ldots$

$$(3.14) \qquad \begin{aligned} f(\mathbf{z}\sin\theta, \cos\theta) &= (\mathbf{z}\sin\gamma_k(\theta), \cos\gamma_k(\theta)) \\ \gamma_k(\theta) &= \theta + \frac{2\pi(\theta - \theta_k)}{\theta_k} \end{aligned}$$

Let us observe that $f : \mathbb{S}_{\theta_k}^{2\theta_k} \to \mathbb{S}^n$ is the identity on the boundary of $\mathbb{S}_{\theta_k}^{2\theta_k}$. Precisely, we have $\gamma_k(\theta_k) = \theta_k$ and $\gamma_k(2\theta_k) = 2\theta_k + 2\pi$. Furthermore, f maps all points of latitude $\theta = 3\pi\theta_k/(2\pi + \theta_k)$ into the south pole. It maps points of latitude $\theta = 4\pi\theta_k/(2\pi + \theta_k)$ into the north pole. Outside those latitude spheres f is a local diffeomorphism. Since the image of $\mathbb{S}_{\theta_k}^{2\theta_k}$ covers the whole sphere $\mathbb{S}^n$ we estimate the integral of the Jacobian determinant of f as

$$(3.15) \qquad \int_{\mathbb{S}_{\theta_k}^{2\theta_k}} \mathcal{J}(x,f)\,dx \geqslant 2\omega_n$$

for every $k = 1, 2, \ldots$ We then conclude that the Jacobian is not integrable

$$(3.16) \qquad \int_{\mathbb{S}^n} \mathcal{J}(x,f)\,dx \geqslant \sum_{k=1}^{\infty} 2\omega_n = \infty$$

Because of this f does not belong to the Sobolev class $\mathscr{W}^{1,n}(\mathbb{S}^n, \mathbb{S}^n)$. It is desirable to see which Orlicz-Sobolev classes contain this mapping f. We therefore need to estimate the differential of f. On each spherical slice $\mathbb{S}_{2\theta_{k+1}}^{\theta_k}$ the norm of Df equals 1, whereas for $(\mathbf{z}\cos\theta, \sin\theta)$ in $\mathbb{S}_{\theta_k}^{2\theta_k}$ formula (3.11) yields

$$(3.17) \qquad \begin{aligned} |Df(\mathbf{z},\theta)| &= \max\left\{1 + \frac{2\pi}{\theta_k}, \left|\frac{\sin\gamma_k(\theta)}{\sin\theta}\right|\right\} \\ &\leqslant \max\left\{1 + \frac{2\pi}{\theta_k}, \frac{1}{\sin\theta}\right\} \leqslant \frac{8}{\theta_k}, \end{aligned}$$

because $\theta_k \leqslant \theta < 2\theta_k$. Also observe that the volume of $\mathbb{S}_{\theta_k}^{2\theta_k}$ does not exceed $2^n n^{-1}\omega_{n-1}\theta_k^n$, by the formula at (3.8).

Now let $P : [0, \infty) \to [0, \infty)$ be any Orlicz function that exhibits slower growth than t^n. Precisely, we assume that

$$\liminf_{t \to \infty} t^{-n} P(t) = 0 \tag{3.18}$$

The reader may consult Section 4.3 for the definition of the Orlicz function. We find that

$$\int_{\mathbb{S}^{2\theta_k}_{\theta_k}} P(|Df(x)|) \, dx \leqslant |\mathbb{S}^{2\theta_k}_{\theta_k}| \, P(\tfrac{8}{\theta_k}) \leqslant \tfrac{2^n \omega_{n-1}}{n} P(\tfrac{8}{\theta_k}) \tag{3.19}$$

As $|Df(x)| = 1$ on the remaining spherical slices, it follows that

$$\int_{\mathbb{S}^n} P(|Df(x)|) \, dx \leqslant \omega_n \, P(1) + \sum_{k=1}^{\infty} \lambda(\theta_k) \tag{3.20}$$

where

$$\lambda(\epsilon) = \tfrac{2^n \omega_{n-1}}{n} \, \epsilon^n P(\tfrac{8}{\epsilon}) \tag{3.21}$$

Since

$$\liminf_{\epsilon \to 0} \lambda(\epsilon) = 0,$$

one can find a sequence of latitude angles satisfying (3.12) such that

$$\sum_{k=1}^{\infty} \lambda(\theta_k) < \infty$$

We then conclude this section with our primary result.

THEOREM 3.1. *For every Orlicz function satisfying (3.18) there exists an orientation preserving mapping $f : \mathbb{S}^n \to \mathbb{S}^n$ in the Orlicz-Sobolev class $\mathscr{W}^{1,P}(\mathbb{S}^n, \mathbb{S}^n)$ whose Jacobian determinant is not integrable.*

CHAPTER 4

Some Classes of Functions

We recall the Riemannian volume element dx on $\mathbb{X}$. However, our considerations in this section pertain to abstract measure spaces. Perhaps the reader has already observed that we have reserved capital script letters for all types of function spaces, with few exceptions. Thus, let $(\mathbb{X}, dx)$ be a finite measure space and $0 < p < \infty$. The Lebesgue $\mathscr{L}^p$-space, denoted by $\mathscr{L}^p(\mathbb{X})$, is a complete metric space. The metric is induced by the non-linear functional

$$(4.1) \qquad \|F\|_p = \|F\|_{\mathscr{L}^p(\mathbb{X})} = \left(\int_{\mathbb{X}} |F(x)|^p \, dx \right)^{\frac{1}{p}} < \infty$$

see Section 4.3 and formula (4.29).

Before making generalizations we single out the weak-$\mathscr{L}^p$ space, which is also known as Marcinkiewicz space.

4.1. Marcinkiewicz space $\mathscr{L}^p_{\mathrm{weak}}(\mathbb{X})$

This space consists of functions satisfying

$$(4.2) \qquad [F]_p \stackrel{\mathrm{def}}{=} \sup_{t \geqslant 0} \left(t^p \int_{|F|>t} dx \right)^{\frac{1}{p}} < \infty$$

Clearly, we have

$$(4.3) \qquad [F]_p \leqslant \|F\|_p, \quad \text{hence} \quad \mathscr{L}^p(\mathbb{X}) \subset \mathscr{L}^p_{\mathrm{weak}}(\mathbb{X})$$

It is evident that $[\ ]_p$ is not a norm, and also $\|\ \|_p$ fails to be a norm in $\mathscr{L}^p(\mathbb{X})$ when $0 < p < 1$. For every $0 \leqslant \alpha < p$, Fubini's theorem yields

$$\begin{aligned}
t^p \int_{|F|>t} dx &\leqslant t^{p-\alpha} \int_{|F|>t} |F(x)|^\alpha \, dx \\
&= t^p \int_{|F|>t} dx + t^{p-\alpha} \int_{|F|>t} (|F(x)|^\alpha - t^\alpha) \, dx \\
&\leqslant [F]_p^p + t^{p-\alpha} \int_{|F|>t} \left(\int_t^{|F(x)|} \alpha \tau^{\alpha-1} \right) d\tau \, dx \\
&= [F]_p^p + t^{p-\alpha} \int_t^\infty \alpha \tau^{\alpha-p-1} \left(\tau^p \int_{|F|>\tau} dx \right) d\tau
\end{aligned}$$

$$(4.4) \qquad \leqslant \frac{p}{p-\alpha} [F]_p^p$$

Taking the supremum over $t \geqslant 0$, we obtain

$$(4.5) \qquad [F]_p \leqslant [F]_{\alpha,p} \stackrel{\mathrm{def}}{=} \sup_{t \geqslant 0} \left(t^{p-\alpha} \int_{|F|>t} |F|^\alpha \right)^{\frac{1}{p}} \leqslant \sqrt[p]{\frac{p}{p-\alpha}} \, [F]_p$$

In other words, $\mathscr{L}^p_{\text{weak}}(\mathbb{X})$ is characterized by the inequality $[F]_{\alpha,p} < \infty$ for some or (equivalently) for all $0 \leqslant \alpha < p$. The Lebesgue space $\mathscr{L}^p(\mathbb{X})$ corresponds to $\alpha = p$, in which case $\|F\|_p = [F]_{p,p}$. It follows from the above estimates that $\mathscr{L}^p_{\text{weak}}(\mathbb{X}) \subset \mathscr{L}^\alpha(\mathbb{X})$, for every $0 < \alpha < p$. As a mater of fact, we have the estimate

$$(4.6) \qquad \|F\|_\alpha \preccurlyeq |\mathbb{X}|^{\frac{1}{\alpha}-\frac{1}{p}}[F]_p$$

To this end consider the inequalities

$$\int_{\mathbb{X}} |F|^\alpha \leqslant \int_{|F|\leqslant t} |F(x)|^\alpha\, dx + \int_{|F|>t} |F(x)|^\alpha\, dx$$
$$(4.7) \qquad \leqslant t^\alpha |\mathbb{X}| + \frac{p}{p-\alpha} t^{\alpha-p}[F]_p^p$$

by (4.5). Estimate (4.6) is immediate if we take $t = |\mathbb{X}|^{-1/p}[F]_p$.

We shall now place the Marcinkiewicz class $\mathscr{L}^p_{\text{weak}}(\mathbb{X})$ in a family $\mathscr{L}^{\alpha,p}(\mathbb{X})$ of the so-called very weak Lebesgue spaces, see [30], [13].

4.2. The space $\mathscr{L}^{\alpha,p}(\mathbb{X})$

Given $0 \leqslant \alpha < p$, the space $\mathscr{L}^{\alpha,p}(\mathbb{X})$ consists of measurable functions $F = F(x)$ such that

$$(4.8) \qquad \{F\}_{\alpha,p} \stackrel{\text{def}}{=} \liminf_{t\to\infty} \left(t^{p-\alpha} \int_{|F|>t} |F(x)|^\alpha\, dx \right)^{\frac{1}{p}} < \infty$$

One should be a little cautious because $\mathscr{L}^{\alpha,p}(\mathbb{X})$ is not a linear space, though $F \in \mathscr{L}^{\alpha,p}(\mathbb{X})$ implies $\lambda F \in \mathscr{L}^{\alpha,p}(\mathbb{X})$ for every $\lambda \in \mathbb{R}$. Precisely, $\{\lambda F\}_{\alpha,p} = |\lambda|\{F\}_{\alpha,p}$. It is clear that $\mathscr{L}^p(\mathbb{X}) \subset \mathscr{L}^{\alpha,p}(\mathbb{X})$ for every $0 < \alpha < p$. On the other hand,

$$(4.9) \qquad \mathscr{L}^{\alpha,p}(\mathbb{X}) \nsubseteq \bigcup_{s>\alpha} \mathscr{L}^s(\mathbb{X})$$

see Section 4.5. For $0 \leqslant \alpha \leqslant \beta < p$ we have the following chain of inclusions

$$(4.10) \qquad \mathscr{L}^p_{\text{weak}}(\mathbb{X}) \subset \mathscr{L}^{\beta,p}(\mathbb{X}) \subset \mathscr{L}^{\alpha,p}(\mathbb{X}) \subset \mathscr{L}^p_{\text{Weak}}(\mathbb{X})$$

This latter new space $\mathscr{L}^p_{\text{Weak}}(\mathbb{X}) = \mathscr{L}^{0,p}(\mathbb{X})$ consists of functions satisfying

$$(4.11) \qquad \liminf_{t\to\infty} \left(t^p \int_{|F|>t} dx \right)^{\frac{1}{p}} = \{F\}_{0,p} < \infty$$

The nuance is that we have replaced sup in the definition of $\mathscr{L}^p_{\text{weak}}(\mathbb{X})$ by lim inf. Finally, we introduce the subclass $\mathscr{L}^{\alpha,p}_\circ(\mathbb{X}) \subset \mathscr{L}^{\alpha,p}(\mathbb{X})$, $0 \leqslant \alpha < p$, by requiring that $\{F\}_{\alpha,p} = 0$.

4.2.1. Spherical averages. Let $n-1 < \alpha < n$ and $F \in \mathscr{L}^{\alpha,n}_\circ(\mathbb{X})$. For a given point $x \in \mathbb{X}$, we shall look closely at the expressions

$$(4.12) \qquad F_x(r) = r \left(\fint_{\mathbb{S}(x,r)} |F(y)|^\alpha\, dy \right)^{\frac{1}{\alpha}}$$

as functions defined for almost every $r \in (0, R]$, where R will be a small number. As usually, $\mathbb{S}(x,r) = \partial \mathbb{B}(x,r)$ denotes the geodesic sphere in $\mathbb{X}$ centered at x and with radius r. When applied to $F = |Df|$ these integral expressions represent average stretchings of the deformation $f : \mathbb{X} \to \mathbb{Y}$. One important feature of the space $\mathscr{L}^{\alpha,n}_\circ(\mathbb{X})$ is that $F_x(r)$ assumes arbitrarily small values as $r \to 0$. Precise statement reads as:

4.2. THE SPACE $\mathscr{L}^{\alpha,p}(\mathbb{X})$

PROPOSITION 4.1. *For every $\epsilon > 0$ the set of radii $r \in (0, R]$ such that $F_x(r) \leqslant \epsilon$ has positive linear measure.*

PROOF. Recall that $F \in \mathscr{L}_o^{\alpha,n}(\mathbb{X})$ has the property,

$$(4.13) \qquad \liminf_{t \to \infty} t^{n-\alpha} \int_{|F|>t} |F(x)|^\alpha \, dx = 0$$

Assume, to the contrary, that there exists $\epsilon > 0$ such that

$$(4.14) \qquad F_x(r) = r \left(\fint_{\mathbb{S}(x,r)} |F(y)|^\alpha \, dy \right)^{\frac{1}{\alpha}} > \epsilon$$

for almost every $r \in (0, R]$. For simplicity we assume that $\mathbb{S}(x,r)$ are spheres in $\mathbb{R}^n$ and $x = 0$. The general case reduces to this Euclidean one by using the normal coordinates. In this coordinate system small geodesic spheres centered at x become the Euclidean spheres centered at 0, see [**37**, definition 1.4.4.]. Inequality (4.14) translates into the following estimate

$$(4.15) \qquad \begin{aligned} \frac{\omega_{n-1} \rho^{n-\alpha}}{n-\alpha} &= \omega_{n-1} \int_0^\rho \frac{dr}{r^{\alpha-n+1}} < \frac{1}{\epsilon^\alpha} \int_0^\rho \left(\int_{|x|=r} |F(x)|^\alpha \, dx \right) dr \\ &= \frac{1}{\epsilon^\alpha} \int_{|x| \leqslant \rho} |F(x)|^\alpha \, dx \end{aligned}$$

for every $0 < \rho < R$. We split $|F|$ into two parts, say $|F| = F_1 + F_2$

$$(4.16) \qquad F_1(x) = \begin{cases} |F(x)| & \text{if } |F(x)| \leqslant \frac{\epsilon}{2|x|} \\ 0 & \text{if } |F(x)| > \frac{\epsilon}{2|x|} \end{cases}$$

$$(4.17) \qquad F_2(x) = \begin{cases} 0 & \text{if } |F(x)| \leqslant \frac{\epsilon}{2|x|} \\ |F(x)| & \text{if } |F(x)| > \frac{\epsilon}{2|x|} \end{cases}$$

To simplify writing we introduce the parameter $t = \frac{\epsilon}{2\rho}$ and proceed as follows:

$$\begin{aligned} \frac{\omega_{n-1} \rho^{n-\alpha}}{n-\alpha} &< \frac{1}{\epsilon^\alpha} \int_{|x| \leqslant \rho} |F(x)|^\alpha \, dx \\ &= \frac{1}{\epsilon^\alpha} \int_{|x| \leqslant \rho} |F_1(x)|^\alpha \, dx + \frac{1}{\epsilon^\alpha} \int_{|x| \leqslant \rho} |F_2(x)|^\alpha \, dx \\ &\leqslant \frac{1}{2^\alpha} \int_{|x| \leqslant \rho} \frac{dx}{|x|^\alpha} + \frac{1}{\epsilon^\alpha} \int_{|F(x)|>t} |F(x)|^\alpha \, dx \\ &= \frac{\omega_{n-1} \rho^{n-\alpha}}{2^\alpha (n-\alpha)} + \frac{(2\rho)^{n-\alpha}}{\epsilon^n} t^{n-\alpha} \int_{|F(x)|>t} |F(x)|^\alpha \, dx \end{aligned}$$

The first term is absorbed by the left hand side, so we arrive at the estimate

$$(4.18) \qquad \frac{(2^\alpha - 1)\omega_{n-1}\epsilon^n}{2^n(n-\alpha)} \leqslant t^{n-\alpha} \int_{|F|>t} |F(x)|^\alpha \, dx$$

This estimate is in contradiction with (4.13) once we let $t = \frac{\epsilon}{2\rho}$ go to infinity.

4.2.2. Special sequences. By the definition, every $F \in \mathscr{L}^{\alpha,p}(\mathbb{X})$ admits a sequence $\{t_i\}_{i=1,2,\ldots}$ of positive numbers increasing to infinity such that

$$\text{(4.19)} \qquad \sup_{i \geqslant 1} \left(t_i^{p-\alpha} \int_{|F|>t_i} |F(x)|^\alpha \, dx \right)^{\frac{1}{p}} < \infty$$

If $F \in \mathscr{L}^p_{\text{weak}}(\mathbb{X})$ then every sequence $\{t_i\}_{i=1,2,\ldots}$ has this property. Now, a sequence of positive numbers $\{t_i\}_{i=1,2,\ldots}$ increasing to infinity will be referred to as *special sequence* for F if

$$\text{(4.20)} \qquad \lim_{i \to \infty} \left(t_i^{p-\alpha} \int_{|F|>t_i} |F(x)|^\alpha \, dx \right)^{\frac{1}{p}} = 0$$

Thus the notation $F \in \mathscr{L}^{\alpha,p}_\circ(\mathbb{X})$ simply means that F admits a special sequence. It is not difficult to see that if $F, G \in \mathscr{L}^{\alpha,p}_\circ(\mathbb{X})$ have common special sequence $\{t_i\}$ then the nontrivial linear combination $H = \lambda F + \mu G$ also lies in $\mathscr{L}^{\alpha,p}_\circ(\mathbb{X})$. Indeed, special sequence for H consists of the number $\tau_i = (|\lambda| + |\mu|) t_i$, by the following computation:

$$\left(\tau_i^{p-\alpha} \int_{|H|>\tau_i} |H(x)|^\alpha \, dx \right)^{\frac{1}{p}} \leqslant$$

$$(|\mu| + |\lambda|) \left[\left(t_i^{p-\alpha} \int_{|F|>t_i} |F(x)|^\alpha \, dx \right)^{\frac{1}{p}} + \left(t_i^{p-\alpha} \int_{|G|>t_i} |G(x)|^\alpha \, dx \right)^{\frac{1}{p}} \right]$$

However, the linear structure is lost in $\mathscr{L}^{\alpha,p}_\circ(\mathbb{X})$ because two different functions may not have a common special sequence.

The reader may wish to recall the function $F(x) = |Df(x)|$ defined at (3.2). It belongs to the Marcinkiewicz class $\mathscr{L}^n_{\text{weak}}(\mathbb{X})$, but fails to satisfy the condition $\{F\}_{0,n} = 0$. In Section 8.2 we shall make use of special sequences to introduce the concept of the so-called weak integrals. It will not matter which special sequence we choose, they all yield the same weak integral. It is therefore natural and important to know in which of the function spaces we always have special sequences. Let us begin with the Orlicz classes.

4.3. The Orlicz space $\mathscr{L}^P(\mathbb{X})$

In this section we briefly review basic concepts of the theory of Orlicz spaces. Naturally, it also gives us an opportunity to discuss the notation used in this text.

DEFINITION 4.2. *The term Orlicz function pertains to any infinitely differentiable function* $P : \mathbb{R}_+ \to \mathbb{R}_+$ *which is strictly increasing and satisfies*

$$\text{(4.21)} \qquad P(0) \stackrel{\text{def}}{=} \lim_{t \to 0} P(t) = 0$$

$$\text{(4.22)} \qquad P(\infty) \stackrel{\text{def}}{=} \lim_{t \to \infty} P(t) = \infty$$

DEFINITION 4.3. *The Orlicz space is a collection of measurable functions* $u : \mathbb{X} \to \mathbf{V}$, *such that*

$$\text{(4.23)} \qquad \int_\mathbb{X} P\left(\frac{|u(x)|}{k} \right) dx < \infty$$

for some positive $k = k(u)$. This space will be denoted by $\mathscr{L}^P(\mathbb{X}, \mathbf{V})$.

4.3. THE ORLICZ SPACE $\mathscr{L}^P(\mathbb{X})$

Here and in what follows $\mathbf{V}$ is a finite dimensional normed space. If $\mathbf{V} = \mathbb{R}$ we simply write $\mathscr{L}^P(\mathbb{X})$. Note explicitly that the usual convexity of P will not always be required in this paper. Also, if we fix a basis for $\mathbf{V}$ then $u \in \mathscr{L}^P(\mathbb{X}, \mathbf{V})$ if and only if its coordinate functions with respect to this basis belong to $\mathscr{L}^P(\mathbb{X})$. It is easy to see that $\mathscr{L}^P(\mathbb{X}, \mathbf{V})$ is a linear space. Clearly $\mathscr{L}^P(\mathbb{X}, \mathbf{V})$ is the Orlicz space for $P(t) = t^p$. Of special importance to us will be the Orlicz spaces $\mathscr{L}^P(\mathbb{X}, \mathbf{V})$ which are slightly larger than $\mathscr{L}^p(\mathbb{X})$, $1 \leqslant p < \infty$. More precisely, our standing assumption upon P will be the so-called *divergence condition*

$$(4.24) \qquad \int_1^\infty \frac{P(t)}{t^{p+1}}\, dt = \infty, \quad \text{for instance } P(t) = \frac{t^p}{\log(e+t)}$$

Actually, p will be the dimension of $\mathbb{X}$. We shall see in Section 4.5 that

Under the divergence condition at (4.24) we have the inclusion

$$(4.25) \qquad \mathscr{L}^P(\mathbb{X}) \subset \mathscr{L}_\circ^{\alpha,p}(\mathbb{X}) \quad \text{for all } 0 \leqslant \alpha < p$$

The Orlicz space is equipped with the *Luxemburg functional* (no triangle inequality) defined by

$$(4.26) \qquad \|u\|_P = \inf\left\{k > 0;\ \int_\mathbb{X} P\left(\frac{|u(x)|}{k}\right) dx \leqslant 1\right\}$$

Of course, if a function u vanishes almost everywhere in $\mathbb{X}$ then $\|u\|_P = 0$. Otherwise, the infimum in (4.26) is attained at exactly one value $k = \|u\|_P > 0$. As a note of warning, it can happen that

$$(4.27) \qquad \int_\mathbb{X} P\left(\frac{|u(x)|}{\|u\|_P}\right) dx < 1$$

To see this take $P(t) = e^t - 1$, $\mathbb{X} = (0, 1]$ and $u(x) = -\log(x + x \log^2 x) \geqslant 0$. Indeed, elementary computation reveals that

$$\int_0^1 \left(e^{|u(x)|} - 1\right) dx = \frac{\pi}{2} - 1 < 1, \quad \text{whereas } \|u\|_P = 1$$

because

$$\int_0^1 \left(e^{\frac{|u(x)|}{k}} - 1\right) dx = \infty \quad \text{for all } k < 1$$

On the other hand for $u \not\equiv 0$ we have the equation

$$\int_\mathbb{X} P\left(\frac{|u(x)|}{\|u\|_P}\right) dx = 1, \quad \text{provided } \|u\|_P > 0$$

whenever the defining Orlicz function satisfies a *doubling condition*. This refers to the condition

$$(4.28) \qquad P(2t) \leqslant \Bbbk\, P(t)$$

for some $\Bbbk > 1$ and all $t \geqslant 0$. We call $\Bbbk$ the *doubling constant*.

$\mathscr{L}^P(\mathbb{X}, \mathbf{V})$ is a complete linear metric space in which the distance between u and v is measured as

$$(4.29) \qquad \text{dist}_P[u, v] = \underset{\mathscr{L}^P(\mathbb{X})}{\text{dist}}[u, v] \overset{\text{def}}{=} \inf\left\{\rho > 0;\ \int_\mathbb{X} P\left(\frac{|u(x) - v(x)|}{\rho}\right) dx \leqslant \rho\right\}$$

Clearly, if $\text{dist}_P[u, v] > 0$, then

$$(4.30) \qquad \int_\mathbb{X} P\left(\frac{|u(x) - v(x)|}{\rho_0}\right) dx \leqslant \rho_0 \quad \text{for } \rho_0 = \underset{\mathscr{L}^P(\mathbb{X})}{\text{dist}}[u, v]$$

with the possibility of equality to occur sometimes. However, if P satisfies the doubling condition, then the equality always holds.

The triangle inequality $\operatorname{dist}_P[u,v] \leqslant \operatorname{dist}_P[u,w] + \operatorname{dist}_P[w,v]$ follows directly from the elementary inequality
$$\frac{|a+b|}{\rho_1 + \rho_2} \leqslant \max\left\{\frac{|a|}{\rho_1}, \frac{|b|}{\rho_2}\right\}, \quad \rho_1, \rho_2 > 0$$

We apply it to $a = u - w$ and $b = w - v$. The arguments establishing completeness of the space $\mathscr{L}^P(\mathbb{X})$ are much the same as in the case of the space $\mathscr{L}^p(\mathbb{X})$, with $p \geqslant 1$.

Taking $P(t) = t^p$, where $p > 0$, we recover the well know fact that $\mathscr{L}^p(\mathbb{X}, \mathbf{V})$ is a complete linear metric space with respect to the distance
$$\operatorname*{dist}_{\mathscr{L}^P(\mathbb{X})}[u,v] = \|u-v\|_p^{p/(p+1)}$$

For $p \geqslant 1$ the usual distance function $\|u - v\|_p$ has many advantages, such as homogeneity. Unfortunately, if $0 < p < 1$, this rather nice expression $\|u-v\|_p$ fails to satisfy the triangle inequality; $\mathscr{L}^p(\mathbb{X})$ is not a Banach space.

If a sequence u_j converges to u in $\mathscr{L}^P(\mathbb{X}, \mathbf{V})$ then also $\lim\limits_{j\to\infty} \|u_j - u\|_P = 0$. This follows from the inequality

(4.31) $\qquad \|u - v\|_P \leqslant \operatorname*{dist}_{\mathscr{L}^P(\mathbb{X})}[u,v], \quad \text{provided } \operatorname*{dist}_{\mathscr{L}^P(\mathbb{X})}[u,v] \leqslant 1$

Another useful observation is that if functions $u_j \in \mathscr{L}^P(\mathbb{X}, \mathbf{V})$, are supported in a common set of finite measure and converge uniformly to u then $u \in \mathscr{L}^P(\mathbb{X}, \mathbf{V})$. Moreover,

(4.32) $\qquad \lim\limits_{j\to\infty} \|u_j - u\|_P \leqslant \lim\limits_{j\to\infty} \operatorname*{dist}_{\mathscr{L}^P(\mathbb{X})}[u_j, u] = 0$

PROPOSITION 4.4. *Let $\mathbb{X}$ be a finite measure space. The closure of bounded functions in $\mathscr{L}^P(\mathbb{X}, \mathbf{V})$ is a linear subspace given by*

(4.33) $\qquad \mathscr{L}_\infty^P(\mathbb{X}, \mathbf{V}) = \left\{ u;\ \int_\mathbb{X} P\left(\frac{|u(x)|}{k}\right) dx < \infty,\quad \text{for every } k > 0 \right\}$

In fact we have even more precise result. Given $u \in \mathscr{L}^P(\mathbb{X}, \mathbf{V})$, where $\mathbb{X}$ is a finite measure space, its distance to $\mathscr{L}^\infty(\mathbb{X}, \mathbf{V}) \subset \mathscr{L}^P(\mathbb{X}, \mathbf{V})$ can be computed by the rule

(4.34) $\qquad \operatorname*{dist}_{\mathscr{L}^P(\mathbb{X})}[u, \mathscr{L}^\infty] = \inf\left\{ k > 0;\ \int_\mathbb{X} P\left(\frac{|u(x)|}{k}\right) dx < \infty \right\}$

If the space $\mathbb{X}$ possesses some differentiable structure we find the following corollary.

COROLLARY 4.5. *Let Ω be a bounded open subset of $\mathbb{R}^n$ (or any Riemannian n-manifold). Then the closure of $\mathscr{C}^\infty(\Omega, \mathbf{V}) \cap \mathscr{L}^\infty(\Omega, \mathbf{V})$ equals $\mathscr{L}_\infty^P(\Omega, \mathbf{V})$.*

It is to be noted that, in general, bounded functions are not dense in $\mathscr{L}^P(\mathbb{X}, \mathbf{V})$. An example that illustrates this possibility is furnished by $P(t) = e^t - 1$, which defines the so-called *exponential class*

(4.35) $\qquad \operatorname{Exp}\mathscr{L}(\mathbb{X}, \mathbf{V}) = \left\{ u;\ \int_\mathbb{X} e^{\frac{|u(x)|}{k}} dx < \infty,\ \text{for some } k = k(u) > 0 \right\}$

However, $\mathscr{L}_\infty^P(\mathbb{X}, \mathbf{V}) = \mathscr{L}^P(\mathbb{X}, \mathbf{V})$ if the defining Orlicz function satisfies a doubling condition.

We refer to an Orlicz function P as *Young function* if its second derivative is non-negative, i.e. P is convex. In this case $\mathscr{L}^P(\mathbb{X}, \mathbf{V})$ is a Banach space and $\|\ \|_P$ satisfies also the triangle inequality. That is why we shall use the term Luxemburg norm for $\|\ \|_P$, in the convex case. It compares with the distance function rather nicely:

$$(4.36) \qquad \operatorname*{dist}_{\mathscr{L}^P(\mathbb{X})} [u, v] \leqslant \sqrt{\|u - v\|_P}, \text{ provided } \|u - v\|_P \leqslant 1$$

This follows from the inequality $P(\epsilon t) \leqslant \epsilon P(t)$ for all $0 \leqslant \epsilon \leqslant 1$ and $t \geqslant 0$; a simple consequence of convexity. Inequalities (4.31) and (4.36) imply that

$$(4.37) \qquad \lim_{j \to 0} \|u_j - u\|_P = 0 \quad \text{iff} \quad \lim_{j \to 0} \operatorname*{dist}_{\mathscr{L}^P(\mathbb{X})} [u_j, u] = 0$$

Note that even in this convex case the density of bounded functions (Proposition 4.4) in $\mathscr{L}^P(\mathbb{X}, \mathbf{V})$ still requires the doubling condition (4.28) as the example of exponential class demonstrates.

In many situations, when we speak of the space $\mathscr{L}^P(\mathbb{X}, \mathbf{V})$, we do not need to explicitly specify the defining function $P = P(t)$; only its behavior for large values of t will be significant to us. To effectively handle this case we make the following definition. Given $\Phi \in \mathscr{C}[0, \infty)$ (not necessarily increasing), an Orlicz function $P = P(t)$ is said to be equivalent to Φ if there exists $\lambda > 1$ such that

$$(4.38) \qquad \tfrac{1}{\lambda} P \left(\tfrac{t}{\lambda}\right) \leqslant \Phi(t) \leqslant \lambda P(\lambda t), \qquad \text{for all } t \geqslant 0$$

We write it as $P \approx \Phi$. It is not always guaranteed that a given $\Phi \in \mathscr{C}[0, \infty)$ admits an equivalent Orlicz function. However, if two Orlicz functions P and Q are equivalent via the parameter λ, then

$$(4.39) \qquad \tfrac{1}{\lambda} \operatorname*{dist}_{\mathscr{L}^P(\mathbb{X})} [u, v] \leqslant \operatorname*{dist}_{\mathscr{L}^Q(\mathbb{X})} [u, v] \leqslant \lambda \operatorname*{dist}_{\mathscr{L}^P(\mathbb{X})} [u, v]$$

Hence $\mathscr{L}^P(\mathbb{X}, \mathbf{V}) = \mathscr{L}^Q(\mathbb{X}, \mathbf{V})$, as metric spaces. In particular, two Orlicz functions equivalent to a given Φ yield the same metric space. When $\mathbb{X}$ has finite measure it will suffice to assume that (4.38) holds for only large values of t, in symbols $P \sim \Phi$. Then Φ need not be even defined for all t. In this finite measure case we still have $\mathscr{L}^P(\mathbb{X}, \mathbf{V}) = \mathscr{L}^Q(\mathbb{X}, \mathbf{V})$, whenever $P \sim \Phi$ and $Q \sim \Phi$.

Some Orlicz spaces play special role in the theory of Jacobians. The Zygmund spaces already have standard notation which we want to recall here:

- $\mathscr{L} \log \mathscr{L}(\mathbb{X}) = \mathscr{L}^P(\mathbb{X}), \quad P = t \log(e + t)$
- $\mathscr{L} \log^{-1} \mathscr{L}(\mathbb{X}) = \mathscr{L}^P(\mathbb{X}), \quad P = \frac{t}{\log(e+t)}$
- $\mathscr{L}^p \log^\alpha \mathscr{L}(\mathbb{X}) = \mathscr{L}^P(\mathbb{X}), \quad P \sim t^p \log^\alpha(e+t),\ 0 < p < \infty$ and $\alpha \in \mathbb{R}$
- $\mathscr{L}^p \log \log \mathscr{L}(\mathbb{X}) = \mathscr{L}^P(\mathbb{X}), \quad P \sim t^p \log \log(e^e + t),\ 0 < p < \infty$

4.4. Grand $\mathrm{G}\mathscr{L}^p$-space

Let $1 < p < \infty$, we consider functions $F \in \bigcap_{1 \leqslant s < p} \mathscr{L}^s(\mathbb{X})$ furnished with the norm

$$(4.40) \qquad \|F\|_{p)} = \sup_{0 < \epsilon \leqslant p-1} \left(\epsilon \int_{\mathbb{X}} |F(x)|^{p-\epsilon}\, dx\right)^{\frac{1}{p-\epsilon}} < \infty$$

This gives us a Banach space, denoted by $\mathrm{G}\mathscr{L}^p(\mathbb{X})$, which is even larger than the Marcinkiewicz class

$$(4.41) \qquad \mathscr{L}^p(\mathbb{X}) \subset \mathscr{L}^p_{\mathrm{weak}}(\mathbb{X}) \subset \mathrm{G}\mathscr{L}^p(\mathbb{X})$$

It is important to realize that $\mathscr{L}^p(\mathbb{X})$ is not dense in $\mathrm{G}\mathscr{L}^p(\mathbb{X})$. Its closure consists of functions having "vanishing p-modulus", see [34], [30]

$$\lim_{\epsilon \to 0} \epsilon \int_{\mathbb{X}} |F(x)|^{p-\epsilon}\, dx = 0 \tag{4.42}$$

We denote this space by $\mathrm{V}\mathscr{L}^p(\mathbb{X})$. As before, the function $F(x) = |DF(x)|$ defined at (3.2) lies in $\mathrm{G}\mathscr{L}^n(\mathbb{X})$ but not in $\mathrm{V}\mathscr{L}^n(\mathbb{X})$; a cause for the lack of smooth approximation later on. Let us record two more inclusions (see the next section)

$$\mathrm{G}\mathscr{L}^p(\mathbb{X}) \subset \mathscr{L}^{\alpha,p}(\mathbb{X}) \tag{4.43}$$

$$\mathrm{V}\mathscr{L}^p(\mathbb{X}) \subset \mathscr{L}_\circ^{\alpha,p}(\mathbb{X}) \tag{4.44}$$

for every $0 \leqslant \alpha < p$.

4.5. Relations between spaces

In this section we will show that

$$\mathrm{V}\mathscr{L}^p(\mathbb{X}) \subset \bigcup_P \mathscr{L}^P(\mathbb{X}) = \mathscr{L}_\circ^{\alpha,p}(\mathbb{X}) \quad 0 < \alpha < p \tag{4.45}$$

where the union runs over all Orlicz functions P satisfying

- divergence condition

$$\int_1^\infty \frac{P(t)}{t^{p+1}}\, dt = \infty \tag{4.46}$$

- growth condition

$$[t^{-\alpha} P(t)]' \geqslant 0, \quad \text{for large values of } t \tag{4.47}$$

We would like to remind the reader that the divergence condition (4.46) alone is too weak to guarantee that the functions in $\mathscr{L}^P(\mathbb{X})$ belong to $\mathscr{L}^\alpha(\mathbb{X})$. That is why we impose the additional condition (4.47) to ensure that $\mathscr{L}^P(\mathbb{X}) \subset \mathscr{L}^\alpha(\mathbb{X})$.

It is well known that $\mathscr{L}_{\mathrm{weak}}^p(\mathbb{X}) \subset \bigcap_{s<p} \mathscr{L}^s(\mathbb{X})$. Here we shall demonstrate that the space $\mathscr{L}_\circ^{\alpha,p}(\mathbb{X})$ is not contained in $\bigcap_{s<p} \mathscr{L}^s(\mathbb{X})$. Even more, we will construct a function which lies in $\mathscr{L}^{\alpha,p}(\mathbb{X})$ but not in $\mathscr{L}^s(\mathbb{X})$, for any $s > \alpha$. Of course this example also will show the inclusion $\mathscr{L}^{\beta,p}(\mathbb{X}) \subsetneq \mathscr{L}^{\alpha,p}(\mathbb{X})$, where $\alpha < \beta$, because $\mathscr{L}^{\alpha,p}(\mathbb{X}) \subset \mathscr{L}^\alpha(\mathbb{X})$.

Let us notice that

$$\mathscr{L}_\circ^{\alpha,p}(\mathbb{X}) = \left\{ F \in \mathscr{L}^\alpha(\mathbb{X}) \colon \inf_{t>0} t^{p-\alpha} \int_{|F|>t} |F(x)|^\alpha\, dx = 0 \right\} \tag{4.48}$$

because the infimum is zero if and only if lim inf is zero.

PROPOSITION 4.6. *Suppose that $u \in \mathrm{V}\mathscr{L}^p(\mathbb{X})$. Then $u \in \mathscr{L}_\circ^{\alpha,p}(\mathbb{X})$ for all $0 < \alpha < p$.*

PROOF. Given $\epsilon > 0$ we consider the function $\Psi_\epsilon(t) = t^{-\epsilon-1}$ defined for all $t > 0$. Note that

$$\epsilon \int_1^\infty \Psi_\epsilon(t)\, dt = 1 \tag{4.49}$$

Next we make use of Fubini's Theorem:

$$\inf_{t \geq 1} t^{p-\alpha} \int_{|u|>t} |u(x)|^\alpha \, dx \leq \epsilon \int_1^\infty \Psi_\epsilon(t) \, t^{p-\alpha} \int_{|u|>t} |u(x)|^\alpha \, dx \, dt$$

$$= \epsilon \int_{|u|>1} |u(x)|^\alpha \int_1^{|u(x)|} t^{p-\alpha-\epsilon-1} \, dt \, dx$$

(4.50)
$$\leq \frac{\epsilon}{p-\alpha-\epsilon} \int_{\mathbb{X}} |u(x)|^{p-\epsilon} \, dx$$

Letting ϵ go to zero, we find that $u \in \mathscr{L}_\circ^{\alpha,p}(\mathbb{X})$, by (4.48). □

PROPOSITION 4.7. *Given an Orlicz-function P satisfying (4.46) and (4.47), $0 < \alpha < p$, let $u : \mathbb{X} \to \mathbb{R}$ be a measurable function such that*

(4.51)
$$\int_\mathbb{X} P(|u(x)|) \, dx < \infty$$

Then

(4.52)
$$u \in \mathscr{L}_\circ^{\alpha,p}(\mathbb{X})$$

PROOF. First we observe that the condition (4.47) implies $u \in \mathscr{L}^\alpha(\mathbb{X})$. Consider the non-negative function $\Psi = t^{\alpha-p}[t^{-\alpha} P(t)]'$, with large values of t; say $t \geq A$. Then

$$\int_A^\infty \Psi(t) \, dt = \left. \frac{P(t)}{t^p} \right|_A^\infty + (p-\alpha) \int_A^\infty \frac{P(t)}{t^{p+1}} \, dt$$

(4.53)
$$\geq -\frac{P(A)}{A^p} + (p-\alpha) \int_A^\infty \frac{P(t)}{t^{p+1}} \, dt = \infty$$

Next we pick up $T > A$, and compute by using Fubini's Theorem:

$$\left(\int_A^T \Psi(t) \, dt \right) \inf_{t \geq A} t^{p-\alpha} \int_{|u|>t} |u(x)|^\alpha \, dx \leq \int_A^T \Psi(t) \, t^{p-\alpha} \left(\int_{|u|>t} |u|^\alpha \right) dt$$

$$\leq \int_{|u|>A} |u|^\alpha \int_A^{|u|} \Psi(t) t^{p-\alpha} \, dt$$

(4.54)
$$\leq \int_{|u|>A} P(|u|) \leq \int_\mathbb{X} P(|u|)$$

Letting T tend to infinity, the claim follows from (4.51), (4.53) and (4.48). □

PROPOSITION 4.8. *Suppose that $u \in \mathscr{L}_\circ^{\alpha,p}(\mathbb{X})$. Then there exists an Orlicz function P satisfying conditions (4.46) and (4.47) such that*

$$u \in \mathscr{L}^P(\mathbb{X})$$

PROOF. We find a special sequence $\{t_k\}$ such that

$$1 < t_1 < t_1 + 1 < t_2 < t_2 + 1 < t_3 < \ldots$$

and

$$t_k^{p-\alpha} \int_{|u|>t_k} |u(x)|^\alpha \, dx \leq 2^{-k}$$

For each $k = 1, 2, \ldots$ we find a smooth nonnegative bump function η_k on $(0, \infty)$ with support in $(t_k, t_k + 1)$, such that

$$\int_{t_k}^{t_k+1} \eta_k(s) \, ds = (p-\alpha)(t_k+1)^{p-\alpha}$$

We set
$$\eta = \sum_{k=1}^{\infty} \eta_k$$
and
$$P(t) = t^\alpha \int_0^t \eta(s)\, ds, \qquad t \in [0, \infty)$$
Then
$$\int_0^\infty \frac{P(t)}{t^{p+1}}\, dt = \int_0^\infty t^{\alpha-p-1} \frac{P(t)}{t^\alpha}\, dt = \int_0^\infty t^{\alpha-p-1} \left(\int_0^t \eta(s)\, ds \right) dt$$
$$= \int_0^\infty \left(\int_s^\infty t^{\alpha-p-1} \eta(s)\, dt \right) ds = (p-\alpha)^{-1} \int_0^\infty s^{\alpha-p} \eta(s)\, ds$$
$$= (p-\alpha)^{-1} \sum_{k=1}^\infty \int_{t_k}^{t_k+1} s^{\alpha-p} \eta_k(s)\, ds \geq \sum_{k=1}^\infty 1 = \infty$$

On the other hand,
$$\int_{\mathbb{X}} P(|u(x)|)\, dx = \int_{\mathbb{X}} |u(x)|^\alpha \left(\int_0^{|u(x)|} \eta(s)\, ds \right) dx$$
$$= \int_0^\infty \eta(s) \left(\int_{|u|>s} |u(x)|^\alpha\, dx \right) ds$$
$$= \sum_{k=1}^\infty \int_{t_k}^{t_k+1} \eta_k(s) \left(\int_{|u|>s} |u(x)|^\alpha\, dx \right) ds$$
$$\leq \sum_{k=1}^\infty \left(\int_{|u|>t_k} |u(x)|^\alpha\, dx \right) \int_{t_k}^{t_k+1} \eta_k(s)\, ds$$
$$\leq C \sum_{k=1}^\infty t_k^{p-\alpha} \int_{|u|>t_k} |u(x)|^\alpha\, dx$$
$$\leq C 2^{-k}$$
which shows that $u \in \mathscr{L}^P(\mathbb{X})$. $\square$

EXAMPLE 4.9. We prove, be means of an example, that the inclusion in (4.45) is proper, which will also validate the claim at (4.9). There exists a function $F : \mathbb{I} = [0,1] \to [0,\infty)$ in the space $\mathscr{L}_\circ^{\alpha,p}(\mathbb{I})$ such that $F \notin \bigcup_{s>\alpha} \mathscr{L}^s(\mathbb{I})$. It then follows that $F \notin \mathrm{V}\mathscr{L}^p(\mathbb{I})$.

Construction. We define $I_1 = 0$, $a_1 = 0$, $t_1 = 1$ and then by induction:
$$t_k = 2^{k^k}$$
$$I_k = t_k^{-\alpha} t_{k-1}^{\alpha-p} 2^{-k} \qquad k = 2, 3, \ldots,$$
$$a_k = a_{k-1} + I_k$$

Consider the intervals
$$\mathbb{I}_k = [a_{k-1}, a_k), \qquad k = 2, 3, \ldots$$
and denote their length by
$$I_k = a_k - a_{k-1} = |\mathbb{I}_k|$$

Then we set
$$F(x) = \sum_{k=2}^{\infty} t_k \chi_{\mathbb{I}_k}(x) \qquad x \in \mathbb{I} = [0,1].$$

Notice the inequality
$$I_k \leqslant 2^{-k}$$
and thus $\bigcup_k \mathbb{I}_k \subset \mathbb{I}$. First we show that F belongs to $\mathscr{L}_\circ^{\alpha,p}(\mathbb{I})$. The sequence $\{t_k\}_{k=1}^{\infty}$ is special in the sense described in Section 4.2.2. Indeed, for $i > k \geqslant 2$
$$t_k^{p-\alpha} I_i t_i^\alpha = t_k^{p-\alpha} t_{i-1}^{\alpha-p} 2^{-i} \leqslant 2^{-i}$$

Hence

(4.55) $\qquad t_k^{p-\alpha} \int_{F > t_k} |F(x)|^\alpha \, dx = t_k^{p-\alpha} \sum_{i=k+1}^{\infty} I_i t_i^\alpha \leqslant \sum_{i=k+1}^{\infty} 2^{-i} = 2^{-k} \to 0$

as k goes to infinity. Now, we show that F does not lie in $\bigcup_{s > \alpha} \mathscr{L}^s(\mathbb{I})$. Indeed,

(4.56) $\qquad \int_0^1 |F(x)|^s \, dx = \sum_{k=1}^{\infty} t_k^s I_k = \sum_{k=1}^{\infty} 2^{-k} t_k^{s-\alpha} t_{k-1}^{\alpha-p}$

Using elementary inequality
$$\log_2 t_k^s I_k = (s-\alpha) k^k - (p-\alpha)(k-1)^{k-1} - k$$
$$\geqslant (s-\alpha) k^k - (p-\alpha) k^{k-1} - k^{k-1}$$
$$= \Big((s-\alpha) k - (p-\alpha+1)\Big) k^{k-1} \to \infty$$

we obtain
$$\int_0^1 |F(x)|^s \, dx = \infty$$

as desired.

4.6. Sobolev classes

A Sobolev mapping $f : \mathbb{X} \to \mathbb{R}^N$ is a vector field $f = (f^1, f^2, ..., f^N)$ whose coordinate functions lay in the usual Sobolev space $\mathscr{W}^{1,p}(\mathbb{X})$, $1 \leqslant p \leqslant \infty$. As for the mappings $f : \mathbb{X} \to \mathbb{Y}$ between manifolds it has been increasingly acknowledged that the introduction of the Riemannian structure on both $\mathbb{X}$ and $\mathbb{Y}$ is necessary to build a viable theory. Broadly speaking the presence of this additional structure involves no loss of generality and at the same time it pays off handsomely in geometric insights. Adopting the imbedding theorem of J. Nash simplifies matters substantially. Thus we assume that the target manifold $\mathbb{Y}$ is $\mathscr{C}^\infty$-isometrically imbedded in some Euclidean space $\mathbb{R}^N$. Now the term Sobolev mapping $f : \mathbb{X} \to \mathbb{Y}$ pertains to a measurable function $f : \mathbb{X} \to \mathbb{R}^N$ such that $f(x) \in \mathbb{Y}$ for a.e. $x \in \mathbb{X}$.

4.6.1. The Orlicz-Sobolev space $\mathscr{W}^{1,P}(\mathbb{X}, \mathbb{Y})$. This space consists of weakly differentiable functions $f : \mathbb{X} \to \mathbb{Y} \subset \mathbb{R}^N$ such that

(4.57) $\qquad \|f\|_{1,P} = \int_{\mathbb{X}} |f(x)| \, dx + \|Df\|_P < \infty$

We should stress that in order to speak of weakly differentiable mapping $f : \mathbb{X} \to \mathbb{Y}$ we must ensure that $|Df| \in \mathscr{L}^1(\mathbb{X})$. This requires that $P(t)$ grows at least linearly, in symbols $P(t) \succcurlyeq t$. With this assumption in place $\mathscr{W}^{1,P}(\mathbb{X}, \mathbb{Y})$ becomes a complete metric space with respect to the distance

(4.58) $\qquad \operatorname{dist}_{1,P}[f, g] = \|f - g\|_1 + \operatorname{dist}_P[Df, Dg]$

Of course, the distance function depends on the imbedding $\mathbb{Y} \subset \mathbb{R}^N$ though different imbeddings yield the same topology in $\mathscr{W}^{1,P}(\mathbb{X}, \mathbb{Y})$. The weak topology in the linear spaces $\mathscr{L}^P(\mathbb{X}, \mathbb{R}^N)$ makes no sense in the nonlinear class $\mathscr{L}^P(\mathbb{X}, \mathbb{Y})$. But we can speak of weak convergence in $\mathscr{W}^{1,P}(\mathbb{X}, \mathbb{Y})$. In what follows we will be interested in the Orlicz-Sobolev classes $\mathscr{W}^{1,P}(\mathbb{X}, \mathbb{Y})$ such that

$$(4.59) \qquad \int_1^\infty \frac{P(t)\,dt}{t^{n+1}} = \infty$$

Additional conditions, such as (4.28) or (4.47), will also be imposed when necessary.

4.6.2. The Sobolev classes $\mathbf{G}\mathscr{W}^{1,n}(\mathbb{X}, \mathbb{R}^N)$ and $\mathbf{V}\mathscr{W}^{1,n}(\mathbb{X}, \mathbb{R}^N)$. Here the dimension of the domain manifold $\mathbb{X}$ equals $n \geqslant 2$. The grand Sobolev space $\mathrm{G}\mathscr{W}^{1,n}(\mathbb{X}, \mathbb{R}^N)$ consists of the vector functions $f : \mathbb{X} \to \mathbb{R}^N$ such that

$$(4.60) \qquad \|f\|_{1,n)} = \|f\|_1 + \sup_{0 < \epsilon \leqslant n-1} \left(\epsilon \int_\mathbb{X} |Df(x)|^{n-\epsilon}\, dx \right)^{\frac{1}{n-\epsilon}} < \infty$$

This is a norm which makes $\mathrm{G}\mathscr{W}^{1,n}(\mathbb{X}, \mathbb{R}^N)$ a Banach space. The closure of $\mathscr{C}^\infty(\mathbb{X}, \mathbb{R}^N)$ in this space denoted by $\mathrm{V}\mathscr{W}^{1,n}(\mathbb{X}, \mathbb{R}^N)$ is characterized precisely by the condition

$$(4.61) \qquad \lim_{\epsilon \to 0} \epsilon \int_\mathbb{X} |Df(x)|^{n-\epsilon}\, dx = 0$$

However, the density of $\mathscr{C}^\infty(\mathbb{X}, \mathbb{Y})$ in $\mathrm{V}\mathscr{W}^{1,n}(\mathbb{X}, \mathbb{Y})$ will require some work if the target manifold $\mathbb{Y}$ is not a vector space. We shall establish this important fact in Section 5.8.

CHAPTER 5

Smooth Approximation

The first of the main questions that faces us is to whether smooth mappings $f : \mathbb{X} \to \mathbb{Y}$ are dense in the given Sobolev class. As we have said, this approximation problem has already a remarkable history, J. Eells and L. Lemaire [10] first consider $\mathscr{W}^{1,p}(\mathbb{X}, \mathbb{Y})$, with $p > n = \dim \mathbb{X}$. By virtue of the embedding theorem such mappings are continuous. A general fact is that whenever $f : \mathbb{X} \to \mathbb{Y}$ happens to be continuous the usual mollification followed by the projection of a tubular neighborhood of $\mathbb{Y}$ gives the desired approximation of f, see also [6] for the related ideas concerning the space VMO$(\mathbb{X}, \mathbb{Y})$-mappings with vanishing mean oscillations. But the true difficulty shows up below the dimension of $\mathbb{X}$; that is, for $1 \leqslant p < n$. The prevailing idea of our approach is that in the Sobolev classes slightly below $\mathscr{W}^{1,n}(\mathbb{X}, \mathbb{Y})$ we were able to detect certain sets (refered to as webs) on which a given map is still continuous.

5.1. Web like structures

We repeat from Introduction that a web on $\mathbb{X}$ is a compact set $\mathbb{F} \subset \mathbb{X}$ of Lebesgue measure zero whose complement consists of finite number of components $\mathbb{U}_i$, $i = 1, ..., I$ (disjoint open connected sets). We call them meshes of the web; thus,

$$(5.1) \qquad \mathbb{X} \setminus \mathbb{F} = \bigcup_{i=1}^{I} \mathbb{U}_i = \bigcup_{\mathbb{U} \in \mathfrak{W}} \mathbb{U}$$

Let the collection of meshes be denoted by $\mathfrak{W} = \{\mathbb{U}_i;\ i = 1, ..., I\}$. In what follows we will spin webs on $\mathbb{X}$ with arbitrarily small meshes. The precise term for this is

$$(5.2) \qquad \text{fine-diameter } (\mathbb{F}) = \max \{\operatorname{diam} \mathbb{U};\ \mathbb{U} \in \mathfrak{W}\}$$

It seems that in the perspective we need only consider "regular web structures", such as Lipschitz or even more regular. In this paper the web will be no other than a finite union of geodesic spheres in $\mathbb{X}$.

5.2. Vanishing web oscillations

We shall investigate Sobolev mappings $f \in \mathscr{W}^{1,p}(\mathbb{X}, \mathbb{R}^N)$ which have continuous trace along the web. Precisely this means that there exists a continuous function $\varphi : \mathbb{X} \to \mathbb{R}^N$ such that $f - \varphi \in \mathscr{W}^{1,p}_\circ(\mathbb{U}, \mathbb{R}^N)$, for every mesh $\mathbb{U} \in \mathfrak{W}$. Equivalently,

$$(5.3) \qquad f \in \varphi + \mathscr{W}^{1,p}_\circ(\mathbb{X} \setminus \mathbb{F}, \mathbb{R}^N)$$

where $\mathscr{W}^{1,p}_\circ(\mathbb{U}, \mathbb{R}^N)$ is the closure of $\mathscr{C}^\infty_\circ(\mathbb{U}, \mathbb{R}^N)$ in $\mathscr{W}^{1,p}(\mathbb{U}, \mathbb{R}^N)$. Note that the above assumptions imply $\varphi \in \mathscr{W}^{1,p}(\mathbb{X}, \mathbb{R}^N)$. Now, we say that a mapping $f \in \mathscr{W}^{1,p}(\mathbb{X}, \mathbb{R}^N)$ has vanishing web oscillations if to every $\epsilon > 0$ there corresponds a web $\mathbb{F} \subset \mathbb{X}$ such that:

1) fine-diameter$(\mathbb{F}) \leqslant \epsilon$

2) f has continuous trace along $\mathbb{F}$, say $\varphi \in \mathscr{W}^{1,p}(\mathbb{X}, \mathbb{R}^N) \cap \mathscr{C}(\mathbb{X}, \mathbb{R}^N)$
3) For every mesh $\mathbb{U} \in \mathfrak{W}$, it holds

$$\operatorname{osc}(f, \partial \mathbb{U}) \stackrel{\text{def}}{=\!=} \max\{|\varphi(x_1) - \varphi(x_2)|;\ x_1, x_2 \in \partial \mathbb{U}\} < \epsilon$$

Observe that in our definition the Sobolev exponent p plays no role since $\mathbb{F}$ is sufficiently regular. But we shall not make use of this observation.

5.3. Statements of the results

Our first theorem shows that the smooth approximation in $\mathscr{W}^{1,p}(\mathbb{X}, \mathbb{Y})$, $1 \leqslant p \leqslant n$, is still possible for discontinuous mappings provided they have vanishing web oscillations. We shall see later that the vanishing web oscillations always occur in Sobolev spaces slightly weaker than $\mathscr{W}^{1,n}(\mathbb{X}, \mathbb{Y})$.

THEOREM 5.1. *Suppose that $f \in \mathscr{W}^{1,p}(\mathbb{X}, \mathbb{Y})$, $p > 1$, has vanishing web oscillations. Then there exist mappings $f_j \in \mathscr{C}^\infty(\mathbb{X}, \mathbb{Y})$, converging to f in $\mathscr{W}^{1,p}(\mathbb{X}, \mathbb{Y})$, such that*

(5.4) $$|Df_j| \preccurlyeq \mathbf{M}(Df), \text{ almost everywhere in } \mathbb{X}$$

where the implied constant depends only on $\mathbb{X}$ and $\mathbb{Y}$.

Recall that $\mathbf{M}$ is the Hardy-Littlewood maximal operator on $\mathbb{X}$. This theorem, although only auxiliary, will be the key to many more convergence results. Let us emphasize that $\mathbf{M}(Df)$ will possess the same degree of integrability as $|Df|$, hence passing to the limit will be achieved by the Lebesgue Dominated Convergence Theorem.

The longitude projection in Section 3.1 demonstrates that the vanishing web oscillations fail in the Marcinkiewicz class $\mathscr{W}^{1,n}_{\text{weak}}(\mathbb{X}, \mathbb{Y})$. However, the situation is completely different if we assume instead of (3.3) that

(5.5) $$\lim_{t \to \infty} t^n \int_{|Df| > t} dx = 0$$

Such mappings indeed have vanishing web oscillations. As a consequence of Theorem 5.1 we will obtain Theorem 1.1.

It is certainly curious that the vanishing web oscillations occur even under slightly weaker assumptions than (5.5). These weaker assumptions are stated in (1.11). Theorem 1.2 will be a consequence of this observation as well.

This idea applies with great effectiveness to many Orlicz-Sobolev classes $\mathscr{W}^{1,P}(\mathbb{X}, \mathbb{Y})$ in which the defining function $P : [0, \infty) \to [0, \infty)$ satisfies the divergence condition

(5.6) $$\int_1^\infty \frac{P(t)}{t^{n+1}} dt = \infty$$

Here is a typical example of such functions

$$P(t) = \frac{t^n}{\log(e+t) \log \log(e^e + t) \ldots \log .. \log(e^{e^{\cdot^{\cdot}}} + t)}$$

It is probably worth mentioning that the divergence condition at (5.6) is critical for many more phenomena in geometric PDEs. Among them are: the $\mathscr{L}^1$-integrability of Jacobians [34], [13], [42], monotonicity of Sobolev functions [30], and compactness of mappings with finite distortion [31]. For precise statements concerning smooth approximation in $\mathscr{W}^{1,P}(\mathbb{X}, \mathbb{Y})$ we need to impose two additional technical assumptions:

(5.7) $\qquad$ the function $\quad t^{-\alpha}P(t), \quad$ with some $\alpha > n-1, \quad$ is nondecreasing

and the doubling condition

(5.8) $\qquad\qquad P(2t) \leqslant \Bbbk P(t) \quad$ for some $\Bbbk > 1$ and all $t \geqslant 0$

THEOREM 5.2. *Let hypothesis (5.6), (5.7) and (5.8) hold. Then the space $\mathscr{C}^\infty(\mathbb{X}, \mathbb{Y})$ is dense in the metric topology $\mathscr{W}^{1,P}(\mathbb{X}, \mathbb{Y})$.*

We want to emphasize that in most situations these technical assumptions at (5.7) and (5.7) are satisfied.

5.4. Proof of Theorem 5.1

We divide the proof into 5 steps.

5.4.1. Step 1-*Truncations*. Let ϵ be any positive number. We consider a web $\mathbb{F}$ of fine-diameter ϵ such that f has continuous trace φ along $\mathbb{F}$ and

(5.9) $\qquad\qquad \mathrm{osc}\,(f, \partial \mathbb{U}) \leqslant \epsilon, \quad$ for every mesh $\mathbb{U} \in \mathfrak{W}$

Given any mesh $\mathbb{U} \in \mathfrak{W}$, we pick up a point

(5.10) $\qquad\qquad a \in f(\partial \mathbb{U}) \subset \mathbb{Y} \subset \mathbb{R}^N$

and consider a map

(5.11) $\qquad\qquad T_\epsilon \circ f : \mathbb{U} \to \mathbb{R}^N,$

where $T_\epsilon : \mathbb{R}^N \to \mathbb{R}^N$, called truncation operator, is given by

(5.12) $\qquad\qquad T_\epsilon y = a + (y-a)\lambda(|y-a|),$

(5.13) $\qquad\qquad \lambda(t) = \begin{cases} 1 & \text{for } 0 \leqslant t \leqslant 2\epsilon \\ \frac{4(t-\epsilon)\epsilon}{t^2} & \text{for } t \geqslant 2\epsilon \end{cases}$

It is immediate that $0 \leqslant \lambda(t) \leqslant 1$ and

(5.14) $\qquad\qquad |T_\epsilon y - a| \leqslant 4\epsilon \quad$ for every $\;y \in \mathbb{R}^N$

As $T_\epsilon \in \mathscr{C}^1(\mathbb{R}^N, \mathbb{R}^N)$, we see that $T_\epsilon f \in \mathscr{W}^{1,P}(\mathbb{X}, \mathbb{R}^N)$. Using chain rule we compute the Hilbert-Schmidt norm of the $N \times n$-matrix $D(T_\epsilon f) \in \mathbb{R}^{N \times n}$

(5.15) $\qquad |D(T_\epsilon f)|^2 = \lambda^2 |Df|^2 + \left(\frac{2\lambda\lambda'}{|f-a|} + \lambda'\lambda'\right)\left|[D^*f](f-a)\right|^2$

where λ and its derivative λ' are computed at $t = |f(x) - a|$. Recall from algebra that the Hilbert-Schmidt norm of $Df \in \mathbb{R}^{N \times n}$ is given by

(5.16) $\qquad\qquad |Df|^2 = \mathrm{Trace}\,(D^*f\, Df)$

where $D^*f \in \mathbb{R}^{n \times N}$ denotes transpose of Df. Hence $[D^*f](f-a)$ is a vector in $\mathbb{R}^n$. It is important to realize that the last term in (5.15) is non-positive and so we can ignore it to obtain

(5.17) $\qquad\qquad |D(T_\epsilon f)| \leqslant |Df| \quad$ almost everywhere in $\;\mathbb{U}$

We also observe that

(5.18) $\qquad\qquad T_\epsilon f - f \in \mathscr{W}_0^{1,P}(\mathbb{U}, \mathbb{R}^N)$

Indeed, since f has continuous trace along $\mathbb{F}$ there exist $u \in \mathscr{W}_0^{1,P}(\mathbb{U}, \mathbb{R}^N)$ and a continuous mapping $\varphi \in \mathscr{W}^{1,P}(\mathbb{X}, \mathbb{R}^N)$, such that $f = \varphi + u$ on $\mathbb{U}$. We approximate

u by mappings $u_j \in \mathscr{C}_0^\infty(\mathbb{U}, \mathbb{R}^N)$. In view of continuity of the truncation operator $T_\epsilon : \mathscr{W}^{1,p}(\mathbb{U}, \mathbb{R}^N) \to \mathscr{W}^{1,p}(\mathbb{U}, \mathbb{R}^N)$ we conclude that

$$f - T_\epsilon f = f - T_\epsilon[\lim_j(\varphi + u_j)] = \varphi + u - \lim_j T_\epsilon(\varphi + u_j)$$

(5.19)
$$= u + \lim_j[\varphi - T_\epsilon(\varphi + u_j)] \in \mathscr{W}_0^{1,p}(\mathbb{U}, \mathbb{R}^N)$$

This follows from the observation that $\varphi - T_\epsilon(\varphi + u_j)$ vanishes near $\partial \mathbb{U}$. To see this we notice that $\varphi(\partial \mathbb{U})$ lies in the ball $\mathbb{B}(a, \epsilon) \subset \mathbb{R}^N$, by (5.9). Since φ is continuous, the image of a neighborhood of $\partial \mathbb{U}$ lies in $\mathbb{B}(a, 2\epsilon)$. It only remains to notice that T_ϵ in $\mathbb{B}(a, 2\epsilon)$.

As a final step, we perform truncation of f over every mesh $\mathbb{U} \in \mathfrak{W}$ and denote the resulting mapping by $f^\epsilon : \mathbb{X} \to \mathbb{R}^N$. It follows from (5.18) that

(5.20) $$f^\epsilon \in \mathscr{W}^{1,p}(\mathbb{X}, \mathbb{R}^N)$$

and

(5.21) $$|Df^\epsilon(x)| \leq |Df(x)| \quad \text{a.e.} \quad x \in \mathbb{X}$$

It will be important that there is no constant involved in the right hand side. Unfortunately, the image of $\mathbb{X}$ under f^ϵ is not longer in the target manifold $\mathbb{Y}$. However, in making truncation we gain small oscillations. Precisely, we have

(5.22) $$|f^\epsilon(x_1) - f^\epsilon(x_2)| \leq 8\epsilon \quad \text{for all } x_1, x_2 \in \mathbb{U} \in \mathfrak{W}$$

by (5.14). This was true for the original f only when $x_1, x_2 \in \partial \mathbb{U}$

REMARK 5.3. Before leaving this step of the proof, let us remark that one could work with somewhat simpler (though only Lipschitz) truncation operator. However, we prefer the $\mathscr{C}^1$-truncation to the Lipschitz one in order to justify the use of the chain rule.

5.4.2. Step 2.- *Truncations converge in $\mathscr{W}^{1,p}(\mathbb{X}, \mathbb{R}^N)$.* We investigate the limit of f^ϵ as $\epsilon \to 0$. First, by using Poincaré inequality (the version with zero traces) we see that for every $\mathbb{U} \in \mathfrak{W}$

$$\int_\mathbb{U} |f^\epsilon - f|^p \preccurlyeq (\operatorname{diam} \mathbb{U})^p \int_\mathbb{U} |Df^\epsilon - Df|^p$$

(5.23)
$$\preccurlyeq (\operatorname{diam} \mathbb{U})^p \int_\mathbb{U} |Df|^p$$

by (5.17). Since the fine-diameter of $\mathbb{F}$ is no larger than ϵ, we may add those estimates over all meshes, to obtain

(5.24) $$\|f^\epsilon - f\|_{\mathscr{L}^p(\mathbb{X})} \preccurlyeq \epsilon \|Df\|_{\mathscr{L}^p(\mathbb{X})}$$

Hence

(5.25) $$\lim_{\epsilon \to 0} f^\epsilon = f \quad \text{in} \quad \mathscr{L}^p(\mathbb{X}, \mathbb{R}^N)$$

Next, we infer from (5.21) that

(5.26) $$\lim_{\epsilon \to 0} Df^\epsilon = Df, \quad \text{in weak topology of } \mathscr{L}^p(\mathbb{X}, \mathbb{R}^{N \times n})$$

It is at this point important that the estimate at (5.21) involves no constant. Lower semicontinuity of the p-norm yields

(5.27) $$\|Df\|_p \leq \liminf_{\epsilon \to 0} \|Df^\epsilon\|_p \leq \limsup_{\epsilon \to 0} \|Df^\epsilon\|_p \leq \|Df\|_p$$

Hence,

$$\lim_{\epsilon \to 0} \| Df^\epsilon \|_p = \| Df \|_p \tag{5.28}$$

By virtue of uniform convexity of $\mathscr{L}^p(\mathbb{X}, \mathbb{R}^{N \times n})$ we conclude that

$$\lim_{\epsilon \to 0} Df^\epsilon = Df, \quad \text{strongly in } \mathscr{L}^p(\mathbb{X}, \mathbb{R}^{N \times n}) \tag{5.29}$$

as desired.

REMARK 5.4. A fact worth noticing is that f^ϵ is not converging uniformly, unless f is continuous. The reason is that in reality the meshes in the web $\mathfrak{W}$ are significantly smaller than the oscillations of f^ϵ. If they were comparable then the limit mapping would be even Lipschitz continuous.

5.4.3. Step 3.-*Mollification.* The truncated mappings $f^\epsilon : \mathbb{X} \to \mathbb{R}^N$ are not smooth, but they have small local oscillations. We now mollify each f^ϵ, as discussed in Section 2.4. The mollified mappings will be denoted by $f_t^\epsilon \in \mathscr{C}^\infty(\mathbb{X}, \mathbb{R}^N)$, for $0 < t \leqslant t_\mathbb{X}$. The reader may wish to consult (2.47) for the definition of the upper bound $t_\mathbb{X}$. Hence

(i) We have convergence

$$\lim_{t \to 0} \| f_t^\epsilon - f^\epsilon \|_{\mathscr{W}^{1,p}(\mathbb{X}, \mathbb{R}^N)} = 0 \tag{5.30}$$

(ii) It follows by (2.58) and by (5.21) again, that

$$|Df_t^\epsilon| \preccurlyeq \mathbf{M}(Df^\epsilon) \leqslant \mathbf{M}(Df), \quad \text{for all } 0 < t \leqslant t_\mathbb{X} \tag{5.31}$$

Given small $\epsilon > 0$ we shall restrict the mollifying parameter $0 < t \leqslant t_\mathbb{X}$ to an interval $0 < t \leqslant t_\epsilon$. The upper bound t_ϵ is determined by requiring the following:

(iii) For every mesh $\mathbb{U} \in \mathfrak{M}$ and $0 < t \leqslant t_\epsilon$ it holds

$$\underset{\mathbb{U}}{\operatorname{osc}} f_t^\epsilon \preccurlyeq \underset{\mathbb{U}_{t'}}{\operatorname{ess\,osc}} f^\epsilon \leqslant 24\epsilon \quad \text{where } t \preccurlyeq t' \preccurlyeq t \tag{5.32}$$

see formula (vii) of Section 2.4.

The reasoning for the last inequality is as follows. Once t is sufficiently small so is t'. We can choose it small enough to ensure that every $\mathbb{U}_{t'}$ intersects only those meshes of the web which touch $\mathbb{U}$. Then, by triangle inequality, we see that $\underset{\mathbb{U}}{\operatorname{ess\,osc}}[f^\epsilon] \leqslant 3 \cdot 8\epsilon$, because of (5.22).

5.4.4. Step 4.-*Convergence of the mollified truncations.* It is immediate from (5.25), (5.26) and (5.30) that

$$\lim_{\epsilon \to 0} f_{t_\epsilon}^\epsilon = f \quad \text{in } \mathscr{W}^{1,p}(\mathbb{X}, \mathbb{R}^N) \tag{5.33}$$

We also infer from (5.31) that

$$\left| Df_{t_\epsilon}^\epsilon \right| \preccurlyeq \mathbf{M}(Df) \tag{5.34}$$

5.4.5. Step 5.-*Projection onto* $\mathbb{Y}$. In the final step we project f_t^ϵ smoothly onto $\mathbb{Y}$. The actual calculation is reduced to a tubular neighborhood of $\mathbb{Y}$ of sufficiently small width; say

$$\mathbb{Y}_h = \{ y \in \mathbb{R}^N; \ \operatorname{dist}(y, \mathbb{Y}) < h \} \tag{5.35}$$

Note that the closest point projection

$$\Pi : \mathbb{Y}_h \to \mathbb{Y} \tag{5.36}$$

is a map of class $\mathscr{C}^\infty(\mathbb{Y}_h, \mathbb{Y})$. Now, the approximating sequence $\{f_j\}$ of smooth mappings converging to f is obtained as $f_j = \Pi(f_{t_j}^{\epsilon_j})$, where $\epsilon_j \to 0$ and the mollifying parameters $t_j \to 0$, are chosen accordingly. For the proof of Theorem 5.1 we need only show that the mappings $\Pi\left(f_{t_j}^{\epsilon_j}\right) \in \mathscr{C}^\infty(\mathbb{X}, \mathbb{Y})$ converge to f in $\mathscr{W}^{1,p}(\mathbb{X}, \mathbb{R}^N)$. First notice that each $f_{t_j}^{\epsilon_j}$ maps $\mathbb{X}$ into $\mathbb{Y}_h$ if ϵ_j is sufficiently small. This follows from the inequality (5.32) and the fact that $f^\epsilon(\mathbb{X})$ lies in a small tubular neighborhood of $\mathbb{Y}$. As $j \to \infty$ the mappings $f_{t_j}^{\epsilon_j}$ are arbitrarily close to f^{ϵ_j} at some points in each mesh $\mathbb{U}$, see (5.30). For abbreviation, we let $\Pi'(y)$ stand for the differential of Π at $y \in \mathbb{Y}_h$. The remaining reasoning goes without further explanation.

$$\begin{aligned}
\|f - \Pi f_\epsilon\|_{\mathscr{W}^{1,p}(\mathbb{X},\mathbb{Y})} &= \|\Pi f - \Pi f_\epsilon\|_{\mathscr{W}^{1,p}(\mathbb{X},\mathbb{Y})} \\
&= \|\Pi f - \Pi f_\epsilon\|_p + \|D(\Pi f) - D(\Pi f_\epsilon)\|_p \\
&\preccurlyeq \|f - f_\epsilon\|_p + \|\Pi'(f) \circ Df - \Pi'(f_\epsilon) \circ Df_\epsilon\|_p \\
&\preccurlyeq \|f - f_\epsilon\|_p + \|\Pi'(f_\epsilon) \circ (Df - Df_\epsilon)\|_p \\
&\quad + \|\Pi'(f) - \Pi'(f_\epsilon) \circ Df\|_p \\
&\to 0 + 0 + 0 = 0
\end{aligned}$$
(5.37)

The only explanation we owe concerns the last step where we have made appeal to Dominated Convergence Theorem.

5.5. Spinning a web on $\mathbb{X}$

In this subsection we consider a Sobolev mapping $f : \mathbb{X} \to \mathbb{Y}$ whose differential lies in the very weak Lebesgue space $\mathscr{L}^{\alpha,n}(\mathbb{X}, \mathbb{R}^{N \times N})$, where $n - 1 < \alpha < n$, see formula (4.8) for the definition of $\mathscr{L}^{\alpha,n}$. Our goal is to build webs on $\mathbb{X}$ which capture arbitrarily small oscillations of f. Precise statement is contained in the following

PROPOSITION 5.5. *Given $\epsilon > 0$, there exists a finite family $\mathfrak{W} = \{\mathbb{U}_\nu;\ \nu = 1, ..., K\}$ of mutually disjoint open sets $\mathbb{U}_\nu \subset \mathbb{X}$, with $\operatorname{diam} \mathbb{U}_\nu \leqslant \epsilon$, whose union $\mathbb{M} = \bigcup_{\nu=1}^K \mathbb{U}_\nu$ has full measure, and there exists a continuous mapping $\varphi : \mathbb{X} \to \mathbb{R}^N$ such that*

(i)
$$f - \varphi \in \mathscr{W}_\circ^{1,\alpha}(\mathbb{U}_\nu, \mathbb{R}^N)$$

(ii)
$$\operatorname*{osc}_{\partial \mathbb{U}_\nu} \varphi \stackrel{\text{def}}{=} \max\{|\varphi(x_1) - \varphi(x_2)|;\ x_1, x_2 \in \mathbb{U}_\nu\} \leqslant \epsilon$$

for all $\nu = 1, 2, ..., K$.

It is automatic from (i) that $\varphi \in \mathscr{W}^{1,\alpha}(\mathbb{X}, \mathbb{R}^N)$. We shall then consider the web $\mathbb{F} = \mathbb{F}_\epsilon = \mathbb{X} - \cup_{\nu=1}^K \mathbb{U}_\nu$. The key ingredient needed for the construction of such webs will be the following

LEMMA 5.6. [OSCILLATIONS ON SPHERES] *Let $h \in \mathscr{C}^\infty(\mathbb{X}, \mathbb{R}^N)$ and $R \leqslant R_\mathbb{X}$ (reliable radius for $\mathbb{X}$ see Section 2.1.1.). Then for every $a \in \mathbb{X}$ and $r \in (0, R]$ we have*

$$\operatorname*{osc}_{\mathbb{S}(a,r)} h \preccurlyeq r \left(\fint_{\mathbb{S}(a,r)} |Dh(x)|^\alpha\, dx \right)^{\frac{1}{\alpha}}$$
(5.38)

provided $\alpha > n - 1$.

This is none other than a spherical variant of the imbedding inequality. That is why the Sobolev exponent α is required to be greater than the dimension of the sphere. We have now all requisites needed for the proof of Proposition 5.5.

PROOF OF PROPOSITION 5.5. Given a sequence $\{f_j\}$ of mappings $f_j \in \mathscr{C}^\infty(\mathbb{X}, \mathbb{R}^N)$, $j = 1, 2, ...$, converging to f in $\mathscr{W}^{1,\alpha}(\mathbb{X}, \mathbb{R}^N)$. Fix a positive number $R \leqslant \min\{\epsilon, R_\mathbb{X}\}$, so that we can use the oscillation Lemma 5.6. Accordingly,

$$
\begin{aligned}
\underset{\mathbb{S}(x,r)}{\operatorname{osc}} g &\leqslant C_\alpha(\mathbb{X}) r \left(\fint_{\mathbb{S}(x,r)} |Dg|^\alpha \right)^{\frac{1}{\alpha}}, \\
\sup_{\mathbb{S}(x,r)} |g| &\leqslant \inf_{\mathbb{S}(x,r)} |g| + \underset{\mathbb{S}(x,r)}{\operatorname{osc}} g \\
&\leqslant \left(\fint_{\mathbb{S}(x,r)} |g|^\alpha \right)^{\frac{1}{\alpha}} + C_\alpha(\mathbb{X}) r \left(\fint_{\mathbb{S}(x,r)} |Dg|^\alpha \right)^{\frac{1}{\alpha}}
\end{aligned}
\tag{5.39}
$$

whenever $0 < r \leqslant R$ and $g\colon \mathbb{X} \to \mathbb{R}^N$ is a smooth function. On the other hand, since $|Df| \in \mathscr{L}^{\alpha,n}(\mathbb{X})$, we may appeal to Proposition 4.1 to ensure the inequalities

$$
C_\alpha(\mathbb{X}) r \left(\fint_{\mathbb{S}(x,r)} |Df|^\alpha \right)^{\frac{1}{\alpha}} \leqslant \frac{\epsilon}{4}
\tag{5.40}
$$

for some radii r in a set of positive linear measure in $(0, R]$. Next, Fubini's theorem tells us that

$$
\lim_{j \to \infty} \int_{\mathbb{S}(x,r)} (|Df_j - Df|^\alpha + |f_j - f|^\alpha) = 0
\tag{5.41}
$$

for almost every r in $(0, R]$. When confronted with (5.40), this gives at least one radius $r = r_x \in (0, R]$ for which (5.40) and (5.41) hold. We consider the covering $\mathbb{X} = \bigcup_{x \in \mathbb{X}} \mathbb{B}(x, r_x)$ by geodesic open balls. Since $\mathbb{X}$ is compact, a finite collection of these balls will also cover $\mathbb{X}$. We assort this finite collection further to obtain a sequence, denoted by $\mathbb{B}_1, ..., \mathbb{B}_k$, $\mathbb{B}_i = \mathbb{B}(x_i, r_i)$, such that

(i)
$$\mathbb{B}_1 \cup ... \cup \mathbb{B}_k = \mathbb{X}$$

(ii) *No ball in the sequence is contained in the other ball.*

Having these selected balls at hand we now define a web $\mathbb{F} = \mathbb{F}_\epsilon$ to be the union of the spheres $\mathbb{S}_i = \partial \mathbb{B}_i$, $i = 1, 2, ..., k$. Then the meshes $\mathbb{U}_1, ..., \mathbb{U}_K$ are the connected components of $\mathbb{X} \setminus \mathbb{F}$. Note (only for a record) that $K \leqslant 2^k$. Now, we pass to a subsequence, labeled again as f_j, such that
(5.42)
$$\|f_j - f_{j-1}\|_{\mathscr{W}^{1,\alpha}(\mathbb{X})} \leqslant 2^{-j}$$

$$
\left(\fint_{\mathbb{S}_i} |f_j - f|^\alpha \right)^{\frac{1}{\alpha}} + C_\alpha(\mathbb{X}) r \left(\fint_{\mathbb{S}_i} |Df_j - Df|^\alpha \right)^{\frac{1}{\alpha}} \leqslant 2^{-j-3} \epsilon, \qquad i = 1, \ldots, k
$$

We define a truncated sequence $\{\varphi_j\}$ by

$$
\begin{aligned}
\varphi_1 &= f_1, \\
\varphi_j - \varphi_{j-1} &= T_{2^{-j}\epsilon}(f_j - f_{j-1}), \qquad j = 2, 3, \ldots
\end{aligned}
$$

where $T_{2^{-j}\epsilon}$ is the truncation operator defined in (5.12) with $2^{-j}\epsilon$ in place of ϵ and with $a = 0$. The properties of the truncation operator ensure that

$$
\sup_{\mathbb{X}} |\varphi_j - \varphi_{j-1}| \leqslant 2^{-j+2} \epsilon
\tag{5.43}
$$

see inequality (5.14). Next we apply (5.39) to the mapping $f_j - f_{j-1}$ in place of g and in view of (5.42) we obtain

(5.44) $$\sup_{\mathbb{S}_i} |f_j - f_{j-1}| \leqslant 2^{-j-2}\epsilon$$

This latter estimate combined with formulas (5.12) and (5.13) show that $T_{2^{-j}\epsilon}(f_j - f_{j-1}) = f_j - f_{j-1}$ on every $\mathbb{S}_i$, $i = 1, ..., k$. In particular, we see that

(5.45) $$\varphi_j = f_j \quad \text{on } \mathbb{F}$$

Appealing to (5.21) we have

(5.46) $$\|D(\varphi_j - \varphi_{j-1})\|_{\mathscr{L}^\alpha(\mathbb{X})} \leqslant \|D(f_j - f_{j-1})\|_{\mathscr{L}^\alpha(\mathbb{X})} \leqslant 2^{-j}$$

By (5.43) and (5.46), the sequence $\{g_j\}$ is a Cauchy sequence in $\mathscr{W}^{1,\alpha}(\mathbb{X})$ and in $\mathscr{C}(\mathbb{X})$. We define φ to be the uniform limit of the sequence $\{\varphi_j\}$. Notice that $\varphi_j - f_j$ are Lipschitz continuous functions on $\mathbb{X}$ and vanish on $\mathbb{F}$, by (5.45). As these functions converge to $\varphi - f$ in $\mathscr{W}^{1,\alpha}(\mathbb{X})$ we deduce that $\varphi - f \in \mathscr{W}_o^{1,\alpha}(\mathbb{U})$ for each connected component $\mathbb{U}$ of $\mathbb{X} \setminus \mathbb{F}$. It remains to estimate the oscillation of φ on the web. We infer from (5.39), (5.40), (5.41) and (5.45) that

(5.47) $$\operatorname*{osc}_{\mathbb{S}_i} \varphi \leqslant \liminf_{j} \operatorname*{osc}_{\mathbb{S}_i} \varphi_j \leqslant C_\alpha(\mathbb{X}) r_i \liminf_{j} \left(\fint_{\mathbb{S}_i} |D\varphi_j|^\alpha \right)^{\frac{1}{\alpha}}$$
$$\leqslant C_\alpha(\mathbb{X}) r_i \left(\fint_{\mathbb{S}_i} |Df|^\alpha \right)^{\frac{1}{\alpha}} \leqslant \frac{\epsilon}{4}$$

Now, each $\mathbb{U}_\nu$ lies in a ball $\mathbb{B}$ from the family $\{\mathbb{B}_1, ..., \mathbb{B}_k\}$. In particular,

(5.48) $$\operatorname{diam} \mathbb{U}_\nu \leqslant \epsilon$$

Further, $\partial \mathbb{U}_\nu$ consists of certain subsets of the spheres $\mathbb{S}_1, ..., \mathbb{S}_k$, only those spheres which intersect $\mathbb{B}$. By condition (ii), every such sphere intersects $\partial \mathbb{B}$. Consequently, given two points $x_1, x_2 \in \partial \mathbb{U}_\nu$, say $x_1 \in \mathbb{S}_{i_1}$ and $x_2 \in \mathbb{S}_{i_2}$, we can find $a_1 \in \mathbb{S}_{i_1} \cap \partial \mathbb{B}$ and $a_2 \in \mathbb{S}_{i_2} \cap \partial \mathbb{B}$ and conclude by triangle inequality that

$$|\varphi(x_1) - \varphi(x_2)| \leqslant |\varphi(x_1) - \varphi(a_1)| + |\varphi(a_1) - \varphi(a_2)|$$
$$+ |\varphi(a_2) - \varphi(x_2)|$$
$$\leqslant \operatorname*{osc}_{\mathbb{S}_{i_1}} \varphi + \operatorname*{osc}_{\mathbb{S}} \varphi + \operatorname*{osc}_{\mathbb{S}_{i_2}} \varphi$$

(5.49) $$\leqslant \epsilon$$

by (5.47). □

5.6. Proof of Theorems 1.1 and 1.2

The distance in the space $\mathscr{W}^{1,n}_{\text{weak}}(\mathbb{X}, \mathbb{Y})$ will be denoted by

(5.50) $$\operatorname{Dist}[f, g] = \|f - g\|_{\mathscr{L}(\mathbb{X})} + \sup_{t \geqslant 0} \left(t^n \int_{|Df - Dg| > t} dx \right)^{\frac{1}{n+1}}$$

It is obvious that (5.5) holds for smooth mappings and remains valid in the closure of $\mathscr{C}^\infty(\mathbb{X}, \mathbb{Y})$. The only non-trivial part of Theorem 1.1 is that every $f \in \mathscr{W}^{1,n}_{\text{weak}}(\mathbb{X}, \mathbb{Y})$ satisfying the condition (1.9) can be approximated by smooth mappings. In view of the inclusion at (4.10) we see that $|Df|$ satisfies (1.10), for every $0 \leqslant \alpha < n$. Proposition 5.5 tells us that f has vanishing web oscillations. Then by virtue of Theorem 5.1, there exist smooth mappings $f_j : \mathbb{X} \to \mathbb{Y}$ converging to f in every $\mathscr{W}^{1,\alpha}(\mathbb{X}, \mathbb{Y})$, $1 \leqslant \alpha < n$. It is important that $|Df_j|$ are dominated, point-wise

almost everywhere, by the maximal function $\mathbf{M}(Df)$; that is, independently of $j = 1, 2, ...$, see (5.4). Since the operator $\mathbf{M} : \mathscr{L}^n_{\text{weak}} \to \mathscr{L}^n_{\text{weak}}$ is bounded, it follows that $\{f_j\}$ is bounded in $\mathscr{W}^{1,n}_{\text{weak}}(\mathbb{X}, \mathbb{Y})$. Lebesgue Dominated Convergence Theorem yields
$$\lim_{j \to \infty} \text{Dist}[f_j, f] = 0$$
completing the proof of Theorem 1.1.

The reader may see from Proposition 5.5 that (1.11) implies vanishing web oscillations. Theorem 1.2 then follows from Theorem 5.1.

5.7. Proof of Theorem 5.2

Given $f \in \mathscr{W}^{1,P}(\mathbb{X}, \mathbb{Y})$ we find a sequence $f_j \in \mathscr{C}^\infty(\mathbb{X}, \mathbb{Y})$ converging to f in $\mathscr{W}^{1,1}(\mathbb{X}, \mathbb{Y})$, whose differentials Df_j are controlled point-wise by the maximal function of Df, as in the inequality (5.4). In particular, $\{Df_j\}$ contains a subsequence converging point-wise almost everywhere to Df. Since $\mathbf{M} : \mathscr{L}^P \to \mathscr{L}^P$ is bounded, again by Lebesgue Dominated Convergence Theorem we conclude that this subsequence converges to f in the metric topology of $\mathscr{W}^{1,P}(\mathbb{X}, \mathbb{Y})$.

5.8. Proof of Theorem 1.3

The distance function in $\mathrm{V}\mathscr{W}^{1,n}(\mathbb{X}, \mathbb{Y})$ is the one induced by the norm at (4.60). This space is contained in every $\mathscr{L}^p(\mathbb{X}, \mathbb{R}^N)$, $1 \leqslant p < n$. Also $\mathbf{M}(Df) \in \mathrm{V}\mathscr{L}^n(\mathbb{X})$. The rest of the proof runs in much the same way as above.

CHAPTER 6

$\mathscr{L}^1$-Estimates of the Jacobian

Let $f : \mathbb{X} \to \mathbb{Y}$ be a Sobolev mapping, where we assume that $n = \dim \mathbb{X} \leqslant \dim \mathbb{Y} = m$. To every $\mathscr{C}^\infty$-smooth n-form $\omega \in \mathscr{C}^\infty(\wedge^n \mathbb{Y})$ there corresponds its pullback $f^\sharp \omega \in \mathscr{L}(\wedge^n \mathbb{X})$ via f. The point-wise estimate

(6.1) $$|f^\sharp \omega| \preccurlyeq |Df|^n$$

gives us at least some idea how to control the degree of integrability of the pullback $f^\sharp \omega$ in terms of $|Df|$. Surprisingly, if $d\omega = 0$, then $f^\sharp \omega$ may enjoy higher degree of integrability than $|Df|^n$. This phenomenon, first observed in [44] and [9] for mappings in $\mathscr{W}^{1,n}(\mathbb{R}^n, \mathbb{R}^n)$, has come to play a central role in modern calculus of variations, nonlinear elasticity and the geometric function theory. Our integral estimates in this paper are sharp generalizations of these results in the manifold setting. If we wish not to make any topological assumption on the target manifold then we need to restrict ourselves to the pullbacks of Cartan n-forms

(6.2) $$\omega = \sum_{i=1}^K \alpha_i \wedge \beta_i, \quad \begin{array}{l} d\alpha_i = d\beta_i = 0 \quad \deg \alpha_i + \deg \beta_i = n \\ \deg \alpha_i \geqslant 1 \quad \text{and} \quad \deg \beta_i \geqslant 1 \end{array}$$

The wedge product structure of the terms will be critical for our arguments. For $f \in \mathscr{W}^{1,n}(\mathbb{X}, \mathbb{Y})$ we have a linear functional that operates on the test function $\varphi \in \mathscr{C}^\infty(\mathbb{X})$ by the rule

$$(f^\sharp \omega)[\varphi] = \int_\mathbb{X} \varphi (f^\sharp \omega) = \int_\mathbb{X} \varphi \sum_{i=1}^K (f^\sharp \alpha_i) \wedge (f^\sharp \beta_i)$$

(6.3) $$\preccurlyeq \|\varphi\|_\infty \int_\mathbb{X} |Df(x)|^n \, dx$$

In other words, $f^\sharp \omega$ can be viewed as a Schwartz distribution of order zero. The differential forms $f^\sharp \alpha_i$ and $f^\sharp \beta_i$ are closed and, therefore, exact modulo harmonic fields. One of the useful analytic advantages of this idea is that $f^\sharp \omega$ can be defined as a Schwartz distribution for all mappings

(6.4) $$f \in \mathscr{W}^{1,s}(\mathbb{X}, \mathbb{Y}), \quad \text{with } s = \frac{n^2}{n+1}$$

see Section 6.1 for details. In particular, if $f_j : \mathbb{X} \to \mathbb{Y}$ are smooth mappings converging to f in $\mathscr{W}^{1,s}(\mathbb{X}, \mathbb{Y})$ then the pullback $f^\sharp \omega$ can be computed by the formula

(6.5) $$(f^\sharp \omega)[\varphi] = \lim_{j \to \infty} \int_\mathbb{X} \varphi (f_j^\sharp \omega)$$

We call $s = \frac{n^2}{n+1}$ the *critical exponent* because this is the smallest one for which we have existence of the limit at (6.5).

In this larger class of mappings the pullback $f^\sharp \omega$ will be a distribution of order 1, meaning that

$$(6.6) \qquad (f^\sharp \omega)[\varphi] \preccurlyeq (\|\varphi\|_\infty + \|d\varphi\|_\infty) \|Df\|^n_{\mathscr{L}^s(\mathbb{X})}$$

for every $\varphi \in \mathscr{C}^\infty(\mathbb{X})$. The computation in the forthcoming section will lead us to integral formulas for the distribution $f^\sharp \omega$. To this effect we shall introduce pointwise Jacobian $\mathcal{J}_\omega(x, f)$

$$(6.7) \qquad \mathcal{J}_\omega(x, f)\, dx = \sum_{i=1}^{K} (f^\sharp \alpha_i) \wedge (f^\sharp \beta_i) = f^\sharp \omega$$

Here we notice that $d(f^\sharp \alpha_i) = 0$ and $d(f^\sharp \beta_i) = 0$. It suggests that we must consider even more general wedge products of closed forms.

6.1. Weak wedge products

Consider closed differential forms $\Phi \in \mathscr{L}^p(\wedge^l \mathbb{X}) \cap \ker d$ and $\Psi \in \mathscr{L}^r(\wedge^k \mathbb{X}) \cap \ker d$, where $1 \leqslant k, l < n$, $k + l = n$ and $1 < p, r < \infty$. First assume that p and r are Hölder conjugate. Thus $\Phi \wedge \Psi$ is integrable. It defines a Schwartz distribution of order zero

$$(6.8) \qquad (\Phi \wedge \Psi)[\eta] = \int_\mathbb{X} \eta\,(\Phi \wedge \Psi) \quad \text{for } \eta \in \mathscr{C}^\infty(\mathbb{X})$$

Now the $\mathscr{L}^p$-cohomology theory proves handy, see Section 2.3.2. Accordingly, every closed form $\Phi \in \mathscr{L}^p(\wedge^l \mathbb{X})$ is exact modulo harmonic fields. Precisely, we can write

$$(6.9) \qquad \Phi = d\varphi + \vartheta$$

where the exact component $d\varphi \in \mathscr{L}^p(\wedge^l \mathbb{X})$ and the harmonic field $\vartheta \in \mathscr{C}^\infty(\wedge^l \mathbb{X})$ are given by

$$(6.10) \qquad \varphi = \mathbf{E}\Phi \quad \text{and} \quad \vartheta = \mathbf{H}\Phi$$

The point is that both $\mathbf{E}$ and $\mathbf{H}$ are bounded linear operators,

$$(6.11) \qquad \mathbf{E} : \mathscr{L}^p(\wedge^l \mathbb{X}) \to \mathscr{W}^{1,p}(\wedge^{l-1} \mathbb{X}), \qquad 1 < p < \infty$$

and

$$(6.12) \qquad \mathbf{H} : \mathscr{L}^p(\wedge^l \mathbb{X}) \to \mathscr{C}^\infty(\mathbb{X}), \qquad 1 \leqslant p < \infty$$

Using the decomposition $\Phi = \vartheta + d\varphi$, we split the integral at (6.8) as

$$(6.13) \qquad \begin{aligned} (\Phi \wedge \Psi)[\eta] &= \int_\mathbb{X} \eta\,(\vartheta \wedge \Psi) + \int_\mathbb{X} \eta\,(d\varphi \wedge \Psi) \\ &= \int_\mathbb{X} \eta\,(\vartheta \wedge \Psi) - \int_\mathbb{X} d\eta \wedge (\varphi \wedge \Psi) \end{aligned}$$

the latter integral converges whenever $\varphi \wedge \Psi$ is integrable. Assume now that $1 \leqslant p, r < \infty$ is a Sobolev conjugate pair; that is,

$$(6.14) \qquad \frac{1}{p} + \frac{1}{r} = 1 + \frac{1}{n}$$

in particular, one of the exponents is less than n, say $1 \leqslant p < n$. Then, by Sobolev imbedding, we find that $\varphi \in \mathscr{L}^{\frac{np}{n-p}}(\wedge^{l-1}\mathbb{X})$. The exponent $\frac{np}{n-p}$ is exactly Hölder conjugate to r, ensuring that $\varphi \wedge \Psi \in \mathscr{L}^1(\wedge^{n-1}\mathbb{X})$.

REMARK 6.1. The reader may wish to argue in much the same way for analogous formula in case when $1 \leqslant r < n$.

We are now ready to make the definition.

DEFINITION 6.2. (DISTRIBUTIONAL WEDGE PRODUCT) Let
$$\Phi \in \mathscr{L}^p(\wedge^\ell \mathbb{X}) \cap \ker d \quad \text{and} \quad \Psi \in \mathscr{L}^r(\wedge^k \mathbb{X}) \cap \ker d$$
where $1 \leqslant p, r < \infty$ are Sobolev conjugate exponents. The distribution $\Phi \wedge \Psi$ operates on the test function $\eta \in \mathscr{C}^\infty(\mathbb{X})$ by the rule

$$(6.15) \qquad (\Phi \wedge \Psi)[\eta] \stackrel{\text{def}}{=} \lim_{j \to \infty} \int_{\mathbb{X}} \eta \, (\Phi_j \wedge \Psi_j)$$

where $\Phi_j \in \mathscr{C}^\infty(\wedge^\ell \mathbb{X})$ and $\Psi_j \in \mathscr{C}^\infty(\wedge^k \mathbb{X})$ are closed forms converging to Φ and Ψ in $\mathscr{L}^p(\wedge^\ell \mathbb{X})$ and $\mathscr{L}^r(\wedge^k \mathbb{X})$, respectively.

REMARK 6.3. For this definition let us recall that closed forms in $\mathscr{C}^\infty(\wedge^l \mathbb{X})$ are dense in $\mathscr{L}^p(\wedge^l \mathbb{X}) \cap \ker d$, see Section 2.3.1. Also notice that the limit at (6.15) does not depend on the choice of the sequences $\{\Phi_j\}$ and $\{\Psi_j\}$.

REMARK 6.4. Our arguments above also show that if $\frac{1}{p} + \frac{1}{r} < 1 + \frac{1}{n}$ then the limit at (6.15) still exists when Φ_j and Ψ_j converge to Φ and Ψ weakly in $\mathscr{L}^p(\wedge^\ell \mathbb{X})$ and $\mathscr{L}^r(\wedge^k \mathbb{X})$, respectively.

It is now obvious how to define the distributional pullback. Suppose we are given a Cartan form as in (6.2). For $f \in \mathscr{W}^{1,s}(\mathbb{X}, \mathbb{Y})$, with the critical exponent $s = \frac{n^2}{n+1}$, we consider the closed forms

$$(6.16) \qquad \Phi_i = f^\sharp \alpha_i \in \mathscr{L}^{s/\ell_i}(\wedge^{\ell_i} \mathbb{X}) \cap \ker d$$

$$(6.17) \qquad \Psi_i = f^\sharp \beta_i \in \mathscr{L}^{s/k_i}(\wedge^{k_i} \mathbb{X}) \cap \ker d$$

where we observe that

$$(6.18) \qquad \frac{\ell_i}{s} + \frac{k_i}{s} = 1 + \frac{1}{n}$$

Thus $\Phi_i \wedge \Psi_i$ can be regarded as a Schwartz distribution. The distributional pullback of ω is then defined by

$$(6.19) \qquad (f^\sharp \omega)[\eta] = \sum_{i=1}^{K} (\Phi_i \wedge \Psi_i)[\eta]$$

6.2. Distributional Jacobian

It is reasonable to ask how the distributional pullback relates to the point-wise Jacobian. The answer is obvious if $f \in \mathscr{W}^{1,n}(\mathbb{X}, \mathbb{Y})$, we simply have $(f^\sharp \omega)[\varphi] = \int_{\mathbb{X}} \varphi(x) \mathcal{J}_\omega(x, f) \, dx$. However one can go slightly below this regularity assumption.

THEOREM 6.5. Let $f : \mathbb{X} \to \mathbb{Y}$, $n = \dim \mathbb{X} \leqslant \dim \mathbb{Y}$, be a Sobolev map satisfying

$$(6.20) \qquad \liminf_{t \to \infty} t^{\frac{n}{n+1}} \int_{|Df| > t} |Df(x)|^{\frac{n^2}{n+1}} \, dx = 0$$

Then there are measurable sets $\mathbb{X}_1 \subset \mathbb{X}_2 \subset \ldots \subset \mathbb{X}$ whose union is $\mathbb{X}$ such that the distributional pullback of every Cartan form $\omega \in \mathscr{C}^\infty(\wedge^n \mathbb{Y})$ takes the form

$$(6.21) \qquad (f^\sharp \omega)[\varphi] = \lim_{j \to \infty} \int_{\mathbb{X}_j} \varphi(x) \mathcal{J}_\omega(x, f) \, dx, \qquad \varphi \in \mathscr{C}^\infty(\mathbb{X})$$

The reader is warned that the sets $\mathbb{X}_1, \mathbb{X}_2, \ldots$ are chosen for a specific map f, the limit at (6.21) may not exist for other sets. As a matter of fact $\mathbb{X}_j$ will be carefully selected from the level sets of the maximal function of $|Df|^{\frac{n^2}{n+1}}$. In particular, the point-wise Jacobian will be bounded on each of those sets, making the integrals and their limit at (6.21) to exist.

At this stage we are able to give meaning to the so-called weak integral of the Jacobian. This is none other than the value of the distribution $f^\sharp \omega$ on the test function identically equal to 1. Formula (6.21) gives

$$(6.22) \qquad (f^\sharp \omega)\,[1] \stackrel{\text{def}}{=\!=} \lim_{j\to\infty} \int_{\mathbb{X}_j} \mathcal{J}_\omega(x,f)\,dx$$

The following corollary is straightforward by a monotone convergence argument.

THEOREM 6.6. *If, in addition to the conditions stated in Theorem 6.5, the Jacobian is nonnegative then it is integrable and coincides with the distribution $f^\sharp \omega$. Precisely, we have*

$$(6.23) \qquad \int_\mathbb{X} \varphi(x) \mathcal{J}_\omega(x,f)\,dx = (f^\sharp \omega)\,[\varphi]$$

for all $\varphi \in \mathscr{C}^\infty(\mathbb{X})$.

Passing to the limit under the integral sign at (6.21) is perfectly justified whenever the point-wise Jacobian is integrable over $\mathbb{X}$. Thus, we also have the following variant of Theorem 6.6.

THEOREM 6.7. *Under the conditions stated in Theorem 6.5, if $\mathcal{J}_\omega(\cdot,f) \in \mathscr{L}^1(\mathbb{X})$, then the point-wise Jacobian coincides with the distribution $f^\sharp \omega$. Precisely, this means that*

$$(6.24) \qquad \int_\mathbb{X} \varphi(x) \mathcal{J}_\omega(x,f)\,dx = (f^\sharp \omega)\,[\varphi]$$

for all $\varphi \in \mathscr{C}^\infty(\mathbb{X})$.

In what follows we refer to Sobolev mappings having non-negative Jacobians $\mathcal{J}_\omega(x,f)$ as orientation preserving. We reserve the notation $\mathcal{J}(x,f)$ for the Jacobian if $\omega = dy$. Theorem 6.6 may fail if ω is not a Cartan form, which is the case of the volume form on the n-sphere $\mathbb{S}^n$, see Theorem 3.1. In the Sobolev class $\mathscr{W}^{1,n}(\mathbb{X},\mathbb{Y})$ the orientation preserving mappings satisfy

$$(6.25) \quad \int_\mathbb{X} \mathcal{J}(x,f) \log\left(e + \frac{\mathcal{J}(x,f)}{\int_\mathbb{X} \mathcal{J}(z,f)\,dz}\right) dx \preccurlyeq \|\mathbf{M}\mathcal{J}(\cdot,f)\|_{\mathscr{L}^1(\mathbb{X})} \preccurlyeq \int_\mathbb{X} |Df|^n$$

This simply means that $\mathcal{J}(\cdot,f)$ belongs to the Zygmund space $\mathscr{L}\log\mathscr{L}(\mathbb{X})$, see S. Müller [44] for the Euclidean case. In our manifold setting, in which the target space is not a rational homology sphere this result will follow from the forthcoming $\mathscr{H}^1$-estimates.

We come now to perhaps the most surprising phenomenon. It is true that if $f \in \mathscr{W}^{1,n}(\mathbb{X},\mathbb{Y})$ and $\mathcal{J}_\omega(x,f) \geqslant 0$, then $\mathcal{J}_\omega$ belongs to $\mathscr{L}\log\mathscr{L}(\mathbb{X})$ regardless whether ω is a Cartan form or not. However, the uniform bound at (6.25) will be lost if ω is not a Cartan form.

6.3. Proof of Theorem 6.5

Although our main objective is to prove Theorem 6.5, the arguments we shall use here can be set in more general context, which might be of independent interest. This more general context consists in replacing the forms $f^\sharp \alpha_i$ and $f^\sharp \beta_i$ at (6.7) by arbitrary closed forms.

6.3.1. An integral estimate of wedge products.

LEMMA 6.8. *Let $\Phi \in \mathscr{C}^\infty(\wedge^\ell \mathbb{X})$ and $\Psi \in \mathscr{C}^\infty(\wedge^k \mathbb{X})$, $k = 1, 2, ..., k + \ell = n$ be closed differential forms and let $p, q \geqslant 1$, satisfy the Sobolev relation $\frac{1}{p} + \frac{1}{q} = 1 + \frac{1}{n}$. Then for every nonempty open set $\Omega \subsetneq \mathbb{X}$ and every test function $\eta \in \mathscr{C}^\infty(\Omega)$, we have*

$$(6.26) \qquad \left| \fint_\Omega \eta (\Phi \wedge \Psi) \right| \preccurlyeq \|\eta\|_{\mathscr{C}^1(\Omega)} \|\mathbf{M}_p \Phi\|_{\mathscr{L}^\infty(\mathbb{X} \smallsetminus \Omega)} \|\mathbf{M}_r \Psi\|_{\mathscr{L}^\infty(\mathbb{X} \smallsetminus \Omega)}$$

Here the implied constant depends only on the manifold $\mathbb{X}$.

PROOF. Because of the relation $\frac{1}{p} + \frac{1}{q} = 1 + \frac{1}{n}$ one of these exponents does not exceed $\frac{2n}{n+1}$. Suppose that

$$(6.27) \qquad 1 \leqslant p \leqslant \frac{2n}{n+1} < n$$

We consider Whitney's covering of Ω by legitimate balls $\mathbb{B}_i$, $i = 1, 2, ...$, as in Proposition 2.1. Next we construct a partition of unity, non-negative functions $\eta_i \in \mathscr{C}_0^\infty(2\mathbb{B}_i)$ whose sum equals 1 on Ω and such that $|d\eta_i| \preccurlyeq (\operatorname{diam} \mathbb{B}_i)^{-1}$ for $i = 1, 2, ...$. Our computation beggins with the formula:

$$(6.28) \qquad \int_\Omega \eta (\Phi \wedge \Psi) = \sum_{i=1}^\infty \int_{2\mathbb{B}_i} \eta_i \eta (\Phi \wedge \Psi)$$

Since $2\mathbb{B}_i$ is a legitimate ball, Poincaré Lemma tells us that the closed form Φ is also exact on $2\mathbb{B}_i$. As a matter of fact, using Sobolev theory of differential forms [35], we find a differential form $\varphi_i \in \mathscr{C}^\infty(\wedge^{\ell-1} 2\mathbb{B}_i)$ whose $\mathscr{W}^{1,p}$-norm is controlled by $\mathscr{L}^p$-norm of Φ, and such that $d\varphi_i = \Phi$. Then, by Sobolev-Poincaré inequality, we obtain

$$(6.29) \qquad \left(\fint_{2\mathbb{B}_i} |\varphi_i|^{\frac{np}{n-p}} \right)^{\frac{n-p}{np}} \preccurlyeq (\operatorname{diam} \mathbb{B}_i) \left(\fint_{2\mathbb{B}_i} |\Phi|^p \right)^{\frac{1}{p}}$$

Next, we integrate (6.28) by parts and use Hölder's inequality with exponents $\frac{np}{n-p}$ and r,

$$\left| \int_{2\mathbb{B}_i} \eta_i \eta (\Phi \wedge \Psi) \right| = \left| \int_{2\mathbb{B}_i} d(\eta_i \eta) (\varphi_i \wedge \Psi) \right|$$

$$\preccurlyeq \|\eta\|_{\mathscr{C}^1(\Omega)} (\operatorname{diam} \mathbb{B}_i)^{-1} \int_{2\mathbb{B}_i} |\varphi_i||\Psi|$$

$$\preccurlyeq \|\eta\|_{\mathscr{C}^1(\Omega)} \frac{|\mathbb{B}_i|}{\operatorname{diam} \mathbb{B}_i} \left(\fint_{2\mathbb{B}_i} |\varphi_i|^{\frac{np}{n-p}} \right)^{\frac{n-p}{np}} \left(\fint_{2\mathbb{B}_i} |\Psi|^r \right)^{\frac{1}{r}}$$

$$(6.30) \qquad \preccurlyeq \|\eta\|_{\mathscr{C}^1(\Omega)} |\mathbb{B}_i| \left(\fint_{2\mathbb{B}_i} |\Phi|^p \right)^{\frac{1}{p}} \left(\fint_{2\mathbb{B}_i} |\Psi|^r \right)^{\frac{1}{r}}$$

At this point it is important to observe that if we enlarge the ball $2\mathbb{B}_i$, by a suitable factor depending only on $\mathbb{X}$, then it will eventually touch the set $\mathbb{X} \smallsetminus \Omega$. This is immediate from the property 4) listed in Proposition 2.1. In other words, there is $\lambda = \lambda(\mathbb{X})$ such that $2\mathbb{B}_i \subset \lambda\mathbb{B}_i$ and $\lambda\mathbb{B}_i \smallsetminus \Omega \neq \emptyset$. We infer from this observation that

$$\text{(6.31)} \qquad \left(\fint_{2\mathbb{B}_i} |\Phi|^p\right)^{\frac{1}{p}} \preccurlyeq \left(\fint_{\lambda\mathbb{B}_i} |\Phi|^p\right)^{\frac{1}{p}} \leqslant \|\mathbf{M}_p\Phi\|_{\mathscr{L}^\infty(\mathbb{X}\smallsetminus\Omega)}$$

Similarly,

$$\text{(6.32)} \qquad \left(\fint_{2\mathbb{B}_i} |\Psi|^p\right)^{\frac{1}{p}} \preccurlyeq \|\mathbf{M}_r\Psi\|_{\mathscr{L}^\infty(\mathbb{X}\smallsetminus\Omega)}$$

Therefore, for each ball $2\mathbb{B}_i$ we can write

$$\text{(6.33)} \qquad \left|\int_{2\mathbb{B}_i} \eta_i\, \eta\, (\Phi \wedge \Psi)\right| \preccurlyeq |\mathbb{B}_i|\, \|\eta\|_{\mathscr{C}^1(\Omega)} \|\mathbf{M}_p\Phi\|_{\mathscr{L}^\infty(\mathbb{X}\smallsetminus\Omega)} \|\mathbf{M}_r\Psi\|_{\mathscr{L}^\infty(\mathbb{X}\smallsetminus\Omega)}$$

As the overlaping number for the covering $\{2\mathbb{B}_i\}_{i=1,2,\ldots}$ depends only on $\mathbb{X}$, we see that $\sum_{i=1}^\infty |\mathbb{B}_i| \preccurlyeq |\Omega|$. Finally, combining (6.28) and (6.33) we conclude with the desired estimate at (6.26).

6.3.2. Point-wise Jacobian versus distributional Jacobian.
Here is the first of our estimates which relates the point-wise Jacobian with distributional Jacobian.

LEMMA 6.9. *Suppose that* $f \in \mathscr{C}^\infty(\mathbb{X}, \mathbb{Y})$ *and* $\lambda > 0$. *Then*

$$\text{(6.34)} \qquad \left|(f^\sharp \omega)[\eta] - \int_{\mathbb{X}\smallsetminus\Omega} \eta(x)\, \mathcal{J}_\omega(x,f)\,dx\right| \preccurlyeq \lambda^n |\Omega|\, \|\eta\|_{\mathscr{C}^1(\mathbb{X})}$$

where $\Omega = \{x;\; (\mathbf{M}_s Df)(x) > \lambda\}$ *and* $s = \frac{n^2}{n+1}$.

PROOF. The left hand side of the inequality (6.34) takes the form

$$\text{(6.35)} \qquad \left|\sum_{i=1}^K \int_\Omega \eta\,(\Phi_i \wedge \Psi_i)\right|$$

where we consider the closed forms $\Phi_i = f^\sharp \alpha_i \in \mathscr{C}^\infty(\wedge^{\ell_i}\mathbb{X})$ and $\Psi_i = f^\sharp \beta_i \in \mathscr{C}^\infty(\wedge^{k_i}\mathbb{X})$. Using Lemma 6.8 we find that

$$\left|\sum_{i=1}^K \int_\Omega \eta\,(\Phi_i \wedge \Psi_i)\right| \leqslant \sum_{i=1}^K \left|\int_\Omega \eta\,(\Phi_i \wedge \Psi_i)\right|$$

$$\text{(6.36)} \qquad \preccurlyeq |\Omega|\, \|\eta\|_{\mathscr{C}^1(\Omega)} \sum_{i=1}^K \|\mathbf{M}_{p_i}\Phi_i\|_{\mathscr{L}^\infty(\mathbb{X}\smallsetminus\Omega)} \|\mathbf{M}_{r_i}\Psi_i\|_{\mathscr{L}^\infty(\mathbb{X}\smallsetminus\Omega)}$$

where $p_i = \frac{s}{\ell_i}$ and $r_i = \frac{s}{k_i}$. Next we observe that

$$\text{(6.37)} \qquad |\Phi_i|^{p_i} = |f^\sharp \alpha_i|^{p_i} \preccurlyeq |Df|^{\ell_i p_i} = |Df|^s$$

and

$$\text{(6.38)} \qquad |\Psi_i|^{r_i} = |f^\sharp \beta_i|^{r_i} \preccurlyeq |Df|^{k_i r_i} = |Df|^s$$

Finally, inequality (6.34) follows readily from the point-wise estimates

$$\text{(6.39)} \qquad \mathbf{M}_{p_i}\Phi_i \preccurlyeq (\mathbf{M}_s Df)^{\ell_i} \leqslant \lambda^{\ell_i}$$

and

$$\text{(6.40)} \qquad \mathbf{M}_{r_i}\Psi_i \preccurlyeq (\mathbf{M}_s Df)^{k_i} \leqslant \lambda^{k_i}$$

where $\ell_i + k_i = n$.

6.3.3. Proof of Theorem 6.5.

First observe that our mapping f also satisfies inequality (6.34) for all but a countable number of parameters $\lambda > 0$. Indeed, by Theorem 1.2 there exist mappings $f_j \in \mathscr{C}^\infty(\mathbb{X}, \mathbb{Y})$ converging to f in $\mathscr{W}^{1,s}(\mathbb{X}, \mathbb{Y})$, $s = \frac{n^2}{n+1}$. We need only justify a passage to the limit in the following inequalities:

$$(6.41) \quad \left|(f_j^\sharp \omega)[\eta] - \int_{\mathbf{M}_s Df_j \leqslant \lambda} \eta(x) \, \mathcal{J}_\omega(x, f_j) \, dx \right| \preccurlyeq \lambda^n \|\eta\|_{\mathscr{C}^1(\mathbb{X})} \int_{\mathbf{M}_s Df_j > \lambda} dx$$

We recall that λ is a regular value of h if the set $\{x;\, h(x) = \lambda\}$ has measure zero. We also point out that the non regular values are always countable. Since, for any $\varepsilon > 0$,

$$(6.42) \quad \int_{\mathbf{M}_s Df_j > \lambda} dx \leqslant \int_{\mathbf{M}_s Df > \lambda - \varepsilon} dx + \int_{\mathbf{M}_s |Df - Df_j| > \varepsilon} dx$$

By the weak-type estimate at (2.52) we obtain

$$(6.43) \quad \limsup_{j \to \infty} \int_{\mathbf{M}_s Df_j > \lambda} dx \leqslant \int_{\mathbf{M}_s Df > \lambda} dx$$

for each regular value λ. Of course, $(f_j^\sharp \omega)[\eta] \to (f^\sharp \omega)[\eta]$, by the definition of the distributional pullback. To deal with the integral in the left hand side we write it as

$$(6.44) \quad \int_\mathbb{X} \eta(x) \, \mathcal{J}_\omega(x, f_j) \chi_j(x) \, dx$$

where χ_j are characteristic functions of the level sets $\{x; \mathbf{M}_s Df_j(x) \leqslant \lambda\}$. Since the integrands are uniformly bounded by $\lambda^n \|\eta\|_{\mathscr{L}^\infty(\mathbb{X})}$, we can apply the Lebesgue Dominated Convergence Theorem. Indeed, $\mathcal{J}_\omega(x, f_j) \to \mathcal{J}_\omega(x, f)$ a.e., so we need only verify that χ_j converge a.e. to χ-the characteristic function of $\{x;\, \mathbf{M}_s Df \leqslant \lambda\}$. This is true for all regular values of $\mathbf{M}_s Df$.

LEMMA 6.10. *Given measurable functions $h_j : \mathbb{X} \to \mathbb{R}$ converging to h almost everywhere, then for every regular value λ*

$$(6.45) \quad \lim_{j \to \infty} \chi_j = \chi \quad a.e.$$

where χ_j and χ are the characteristic functions of the level sets $\{x;\, h_j(x) \leqslant \lambda\}$ and $\{x;\, h(x) \leqslant \lambda\}$, respectively.

We recall that λ is a regular value of h if the set $\{x;\, h(x) = \lambda\}$ has measure zero. We also point out that the set of non regular values is always countable. We can apply this lemma to $h_j = \mathbf{M}_s Df_j$ and $h = \mathbf{M}_s Df$ in view of Corollary 2.15.

Having disposed with the inequality

$$(6.46) \quad \left|(f^\sharp \omega)[\eta] - \int_{\mathbf{M}_s Df \leqslant \lambda} \eta(x) \, \mathcal{J}_\omega(x, f) \, dx \right| \preccurlyeq \lambda^n \|\eta\|_{\mathscr{C}^1(\mathbb{X})} \int_{\mathbf{M}_s Df > \lambda} dx$$

we now fix a special sequence $\{t_j\}$ for $|Df|$. Recall that a special sequence consists of numbers t_j, increasing to infinity, such that

$$(6.47) \quad \lim_{j \to \infty} t_j^{\frac{n}{n+1}} \int_{|Df| > t_j} |Df(x)|^{\frac{n^2}{n+1}} dx = 0$$

There are many such sequences. We choose and fix the one which consists of regular values of $\frac{1}{2}\mathbf{M}_s Df$. Now we are in a position to define the sets $\mathbb{X}_j$

$$(6.48) \quad \mathbb{X}_j = \left\{x;\, (\mathbf{M}_s Df)(x) \leqslant 2t_j\right\} \qquad s = \frac{n^2}{n+1}$$

We make use of the estimate in Lemma 6.9 with $\lambda = 2t_j$; these are regular values of $\mathbf{M}_s Df$.

$$\left| (f^\sharp \omega)[\eta] - \int_{\mathbb{X}_j} \eta(x)\, \mathcal{J}_\omega(x,f)\, dx \right| \preccurlyeq \|\eta\|_{\mathscr{C}^1(\mathbb{X})}\, t_j^n \int_{\mathbf{M}_s Df > 2t_j} dx$$

$$\preccurlyeq \|\eta\|_{\mathscr{C}^1(\mathbb{X})}\, t_j^{n-s} \int_{|Df| > t_j} |Df|^s$$

The latter follows by weak type inequality stated in Proposition (2.14). Letting t_j go to infinity we conclude with (6.21). The proof of Theorem 6.5 is complete.

The interested reader may wish to observe that the above arguments also work for differential forms. Let us state this more general variant of Theorem 6.5 without proof.

THEOREM 6.11. *Given a Cartan form* $\Lambda = \sum_{i=1}^{K} \Phi_i \wedge \Psi_i$, *where* $\Phi_i \in \mathscr{L}^{p_i}(\wedge^{\ell_i}\mathbb{X}) \cap \ker d$ *and* $\Psi_i \in \mathscr{L}^{r_i}(\wedge^{k_i}\mathbb{X}) \cap \ker d$, $1 \leqslant k_i, \ell_i < n$, $k_i + \ell_i = n$. *Here each pair* (p_i, r_i) *consists of Sobolev conjugate exponents. Suppose that*

$$(6.49) \qquad \liminf_{t \to \infty} t^{\frac{1}{n}} \int_{H > t} H(x)\, dx = 0$$

where $H = \sum_{i=1}^{K} \left(|\Phi_i|^{p_i \ell_i} + |\Psi_i|^{r_i k_i} \right)$. *Then there are measurable sets* $\mathbb{X}_1 \subset \mathbb{X}_2 \subset \ldots \subset \mathbb{X}$ *whose union is* $\mathbb{X}$ *such that*

- Λ *is* $\mathscr{L}^1$-*integrable over each* $\mathbb{X}_i$
- *For every* $\eta \in \mathscr{C}^\infty(\mathbb{X})$, *we have*

$$(6.50) \qquad \Lambda[\eta] \stackrel{\mathrm{def}}{=} \sum_{i=1}^{K} (\Phi_i \wedge \Psi_i)[\eta] = \lim_{j \to \infty} \int_{\mathbb{X}_j} \eta \Lambda$$

CHAPTER 7

$\mathscr{H}^1$-Estimates

In this section we formulate and prove the sharpest possible result concerning $\mathscr{H}^1$-regularity of the Jacobian, see [**46**] for somewhat different ideas. Before, we need some auxiliary material.

7.1. The Hausdorff content

Let $s > 0$ and $\mathbb{E} \subset \mathbb{R}^n$. The s-content of $\mathbb{E}$ is defined as

$$(7.1) \qquad \mathfrak{h}^s(\mathbb{E}) = \inf \sum_{j=1}^{\infty} (\operatorname{diam} \mathbb{B}_j)^s,$$

where the infimum is taken over all sequences of balls $\mathbb{B}_j \subset \mathbb{R}^n$ covering the set $\mathbb{E}$.

LEMMA 7.1. *Let* $u \in \mathscr{C}_0^{\infty}(\Omega)$, $\mathbb{B} = \mathbb{B}(a, R) \subset \mathbb{R}^n$ *and* $0 \leqslant n - s < p \leqslant n$. *Then*

$$(7.2) \qquad \mathfrak{h}^s(\{x \in \mathbb{B};\ |u(x)| \geqslant 1\}) \preccurlyeq R^{p+s} \fint_{\mathbb{B}} |\nabla u(x)|^p\, dx$$

For the proof, see [**25**, p. 45]. We infer from this lemma the following useful corollary.

COROLLARY 7.2. *Let* $\frac{n^2}{n+1} < p < n$ *and let* $\mathbb{B} = \mathbb{B}(a, \varrho)$ *be a legitimate ball in* X, $\dim X = n$. *Then every compact set* $\mathbb{E} \subset \mathbb{B}$ *can be split into a finite number of mutually disjoint compact sets* $\mathbb{E}_1, ..., \mathbb{E}_k$ *such that*

$$(7.3) \qquad \sum_{i=1}^{k} \operatorname{diam} \mathbb{E}_i \preccurlyeq \varrho^{n+1} \left(\fint_{\mathbb{B}} |\nabla u(x)|^p\, dx \right)^{\frac{n}{p}}$$

for every test function $u \in \mathscr{C}(\overline{B}) \cap \mathscr{W}^{1,p}(\mathbb{B})$ *satisfying boundary conditions:* $u \leqslant 0$ *on* $\partial \mathbb{B}$ *and* $u \geqslant 1$ *on* $\mathbb{E}$.

PROOF. As the concentric balls $\mathbb{B} \subset 3\mathbb{B}$ lay in a coordinate region Ω, we may change the variables via the diffeomorphism $\kappa : \Omega \xrightarrow{onto} \mathbb{R}^n$, reducing the problem to the Euclidean space. We now apply Lemma 7.1 with $s = \frac{p}{n} < 1$. Clearly, $n - s < p < n$ and

$$(7.4) \qquad \mathfrak{h}^s(\mathbb{E}) \preccurlyeq \varrho^{p+s} \fint_{\mathbb{B}} |\nabla u|^p$$

There exists a finite cover of $\mathbb{E}$ by balls $\mathbb{B}_1, ..., \mathbb{B}_m$ such that

$$(7.5) \qquad \sum_{j=1}^{m} (\operatorname{diam} \mathbb{B}_j)^s \preccurlyeq \varrho^{p+s} \fint_{\mathbb{B}} |\nabla u|^p$$

Let $\mathbb{C}_1, \mathbb{C}_2, ..., \mathbb{C}_k$ be connected components of $\bigcup_{j=1}^{m} \mathbb{B}_j$, and denote $\mathbb{E}_i = \mathbb{E} \cap \mathbb{C}_i$, $i = 1, ..., k$. Obviously, $\mathbb{E}_i$ are mutually disjoint compact sets whose union is $\mathbb{E}$. The rest is the following elementary computation

$$\sum_{i=1}^{k} \operatorname{diam} \mathbb{E}_i \leq \sum_{i=1}^{k} \operatorname{diam} \mathbb{C}_i \leq \sum_{j=1}^{m} \operatorname{diam} \mathbb{B}_j$$

(7.6)
$$\leq \left[\sum_{j=1}^{m} (\operatorname{diam} \mathbb{B}_j)^s\right]^{\frac{1}{s}} \preccurlyeq \varrho^{n+1} \left(\fint_{\mathbb{B}} |\nabla u|^p\right)^{\frac{n}{p}}$$

as claimed. Later we shall choose the following exponents

(7.7) $$p = \frac{2n^2}{2n+1}, \quad \text{so that} \quad s = \frac{2n}{2n+1}$$

7.2. The $\mathscr{H}^1$-Theorem

We shall now state and prove our main result. As regards the dimensions we assume that $2 \leq n = \dim \mathbb{X} \leq \dim \mathbb{Y} = m$. Given any mapping $g \in \mathscr{W}^{1,n}(\mathbb{X}, \mathbb{Y})$. Let $\mathscr{B}(g, \epsilon)$ denote the ball centered at g and with radius ϵ in the space $\mathscr{W}^{1,n}(\mathbb{X}, \mathbb{Y})$, namely

$$\mathscr{B}(g, \rho) = \left\{ f \in \mathscr{W}^{1,n}(\mathbb{X}, \mathbb{Y}) : \|f - g\|_{\mathscr{W}^{1,n}} < \rho \right\}$$

THEOREM 7.3. *Suppose $\omega \in \mathscr{C}^{\infty}(\wedge^n \mathbb{Y})$ is either a Cartan form or a closed form and $g \in \mathscr{W}^{1,n}(\mathbb{X}, \mathbb{Y})$. Then there exist a constant $C(\omega, g)$ and a radius $\epsilon > 0$ such that*

(7.8) $$\|f^{\sharp} \omega\|_{\mathscr{H}^1(\mathbb{X})} \leq C(\omega, g) \int_{\mathbb{X}} |Df(x)|^n \, dx$$

for all $f \in \mathscr{B}(g, \epsilon)$. Moreover, the pullback operator

(7.9) $$\sharp \omega : \mathscr{W}^{1,n}(\mathbb{X}, \mathbb{Y}) \to \mathscr{H}^1(\wedge^n \mathbb{X})$$

is continuous.

REMARK 7.4. If $\omega = \sum \alpha_i \wedge \beta_i$ is a Cartan form, as in (6.2), then in fact

(7.10) $$\|f^{\sharp} \omega\|_{\mathscr{H}^1(\mathbb{X})} \preccurlyeq \|\omega\| \int_{\mathbb{X}} |Df(x)|^n \, dx$$

for all $f \in \mathscr{W}^{1,n}(\mathbb{X}, \mathbb{Y})$ where

(7.11) $$\|\omega\| = \sum \|\alpha_i\|_{\mathscr{L}^{\infty}(\mathbb{Y})} \|\beta_i\|_{\mathscr{L}^{\infty}(\mathbb{Y})}$$

REMARK 7.5. The key to Theorem 7.3 is the following inequality for the Fefferman-Stein maximal function of $f^{\sharp} \omega$, see (2.62) for the definition,

(7.12) $$\mathbf{M}(f^{\sharp} \omega) \preccurlyeq C(\omega, g) \left[\mathbf{M}_p(Df) + \|Df\|_n \right]^n$$

where

(7.13) $$\mathbf{M}_p(Df) = (\mathbf{M}|Df|^p)^{\frac{1}{p}}, \quad p = \frac{2n^2}{2n+1}$$

Indeed, having established the point-wise estimate at (7.12), the inequality at (7.8) becomes straightforward by the maximal theorem. As for the continuity of the pullback $\sharp \omega$, we argue as follows. Let $f_j \to f$ in $\mathscr{W}^{1,n}(\mathbb{X}, \mathbb{Y})$. We need only show that

(7.14) $$\mathbf{M}(f_j^{\sharp} \omega - f^{\sharp} \omega) \to 0 \quad \text{in } \mathscr{L}^1(\mathbb{X})$$

To this end, we may assume that $f_j \in \mathscr{B}(g, \epsilon)$ as this is certainly the case for sufficiently large j. For such j, we have the uniform point-wise inequality

$$\sup_{t>0} \left| \int_{\mathbb{X}} K_t(x, \cdot)(f_j^\sharp \omega - f^\sharp \omega) \right|$$

(7.15) $\quad \preccurlyeq [\mathbf{M}_p(Df_j) + \mathbf{M}_p(Df) + \|Df_j\|_n + \|Df\|_n]^n$

for every $x \in \mathbb{X}$. The right hand side converges in $\mathscr{L}^1(\mathbb{X})$ to $2^n (\mathbf{M}_p(Df) + \|Df\|_n)^n$. It is, therefore, legitimate to use Lebesgue Dominated Convergence Theorem. We shall have established (7.14) if we prove that for almost every $x \in \mathbb{X}$

(7.16) $\quad \sup_{t>0} \left| \int_{\mathbb{X}} K_t(x, \cdot)(f_j^\sharp \omega - f^\sharp \omega) \right| \to 0$

Here the supremum is controlled by means of the Hardy-Littlewood maximal function; we are reduced to proving that

(7.17) $\quad \mathbf{M}(f_j^\sharp \omega - f^\sharp \omega) \to 0 \quad \text{a.e. in } \mathbb{X}$

But this is well known, since $f_j^\sharp \omega \to f^\sharp \omega$ in $\mathscr{L}^1(\mathbb{X})$.

7.2.1. Step 1.-The case of Cartan forms. We begin with the inequality (7.10).

It suffices to consider one wedge product $\omega = \alpha \wedge \beta$, with $\alpha \in \mathscr{C}^\infty(\wedge^l \mathbb{Y}) \cap \ker d$ and $\beta \in \mathscr{C}^\infty(\wedge^{n-l} \mathbb{Y}) \cap \ker d$. We may also assume that $\frac{n}{2} \leqslant l < n$, so that

(7.18) $\quad f^\sharp \omega = f^\sharp \alpha \wedge f^\sharp \beta$

In this formula we have sufficient degree of regularity to ensure that the forms $f^\sharp \alpha$ and $f^\sharp \beta$ are closed. These forms satisfy:

(7.19) $\quad |f^\sharp \alpha| \preccurlyeq \|\alpha\|_\infty |Df|^l \in \mathscr{L}^{\frac{n}{l}}(\mathbb{X})$

(7.20) $\quad |f^\sharp \beta| \preccurlyeq \|\beta\|_\infty |Df|^{n-l} \in \mathscr{L}^{\frac{n}{n-l}}(\mathbb{X})$

The Hodge theory of the deRham cohomology tells us that

(7.21) $\quad f^\sharp \alpha = d\gamma + \chi, \quad \chi \in \mathscr{W}^{1, \frac{n}{l}}(\wedge^{l-1} \mathbb{X})$

where χ is a harmonic field of degree l. It represents the cohomology class of $f^\sharp \alpha$. Harmonic fields, being $\mathscr{C}^\infty$-smooth, are harmless. They form a finite dimensional space and we have nice bounds, such as

(7.22) $\quad \|\chi\|_\infty \preccurlyeq \|\chi\|_s \preccurlyeq \|f^\sharp \alpha\|_s \preccurlyeq \|\alpha\|_\infty \|Df\|_{ls}^l$

provided $1 < s < \infty$. Taking $s = \frac{n}{l}$ we obtain

(7.23) $\quad \|\chi\|_\infty \preccurlyeq \|\alpha\|_\infty \|Df\|_n^l$

Accordingly, we split $f^\sharp \omega$ as

(7.24) $\quad f^\sharp \omega = d\gamma \wedge (f^\sharp \beta) + \chi \wedge (f^\sharp \beta)$

The latter term poses no difficulty as it belongs to $\mathscr{L}^2(\wedge^n \mathbb{X})$. Indeed, we have

$$\|\chi \wedge (f^\sharp \beta)\|_2 \preccurlyeq \|\chi\|_\infty \|f^\sharp \beta\|_2$$
$$\preccurlyeq \|\alpha\|_\infty \|\beta\|_\infty \|Df\|_n^l \||Df|^{n-l}\|_2$$
(7.25) $\quad\quad\quad\quad\quad \preccurlyeq \|\alpha\|_\infty \|\beta\|_\infty \|Df\|_n^n$

Note that we have actually a point-wise estimate of the maximal function

$$\begin{aligned}
\mathcal{M}(\chi \wedge f^\sharp \beta) &\preccurlyeq \|\chi\|_\infty \mathbf{M}(f^\sharp \beta) \\
&\preccurlyeq \|\chi\|_\infty \|\beta\|_\infty \mathbf{M}(|Df|^{n-l}) \\
&\preccurlyeq \|\alpha\|_\infty \|\beta\|_\infty \|Df\|_n^l [\mathbf{M}_p(|Df|)]^{n-l} \\
&\preccurlyeq \|\alpha\|_\infty \|\beta\|_\infty [\|Df\|_n + \mathbf{M}_p(|Df|)]^n
\end{aligned} \tag{7.26}$$

We shall now proceed to the estimates of the maximal function of $d\gamma \wedge (f^\sharp \beta)$. Before jumping to the computation let us observe that the exact l-form $d\gamma$ is not affected by adding any closed form to γ. We have the following variant of the Poincaré-Sobolev inequality for differential forms [35].

$$\left(\int_{\mathbb{B}} |\gamma - \gamma_0|^{\frac{ns}{n-s}} \right)^{\frac{n-s}{ns}} \preccurlyeq \left(\int_{\mathbb{B}} |d\gamma|^s \right)^{\frac{1}{s}}, \quad 1 \leqslant s < n, \tag{7.27}$$

where $\mathbb{B} = \mathbb{B}(x,t)$ is a legitimate ball in $\mathbb{X}$, $0 < t < R_{\mathbb{X}}$, and γ_0 is a suitable closed form on $\mathbb{B}$. As far as integration is concerned we also notice that the mollifying kernels $\zeta \to K_t(x,\zeta)$ are supported in the ball $\mathbb{B} = \mathbb{B}(x,t')$, where t' is comparable with t by a factor depending only on $\mathbb{X}$.

REMARK 7.6. For notational convenience one could introduce new kernel

$$K'_t(x,\zeta) = K_{t'}(x,\zeta)$$

so that the function $\zeta \to K'_t(x,\zeta)$ would be supported in $\mathbb{B} = \mathbb{B}(x,t)$. Instead of doing this we simply assume that $\zeta \to K_t(x,\zeta)$ is supported in $\mathbb{B}(x,t)$.

Integration by parts yields

$$\left| (d\gamma \wedge f^\sharp \beta)_t(x) \right| = \left| \int_{\mathbb{B}} K_t(x,\cdot) d(\gamma - \gamma_0) \wedge f^\sharp \beta \right| \tag{7.28}$$

$$\preccurlyeq \frac{\|\beta\|_\infty}{t^{n+1}} \int_{\mathbb{B}} |\gamma - \gamma_0| |Df|^{n-l}$$

$$\preccurlyeq \frac{\|\beta\|_\infty}{t^{n+1}} \left(\int_{\mathbb{B}} |\gamma - \gamma_0|^{\frac{ns}{n-s}} \right)^{\frac{n-s}{ns}} \left(\int_{\mathbb{B}} |Df|^{\frac{ns(n-l)}{ns-n+s}} \right)^{\frac{ns-n+s}{ns}}$$

We take $s = \frac{n^2}{l(n+1)}$ to obtain, for the center of the ball $\mathbb{B}(x,t)$,

$$\left| (d\gamma \wedge f^\sharp \beta)_t(x) \right| \preccurlyeq \|\beta\|_\infty \left(\fint_{\mathbb{B}} |d\gamma|^s \right)^{\frac{1}{s}} \left(\fint_{\mathbb{B}} |Df|^{\frac{n^2}{n+1}} \right)^{\frac{(n-l)(n+1)}{n^2}} \tag{7.29}$$

Directly from the decomposition $f^\sharp \alpha = d\gamma + \chi$ it follows that

$$\begin{aligned}
|d\gamma| &\leqslant |f^\sharp \alpha| + \|\chi\|_\infty \\
&\preccurlyeq \|\alpha\|_\infty |Df|^l + \|\alpha\|_\infty \|Df\|_n^l
\end{aligned} \tag{7.30}$$

Hence

$$\left(\fint_{\mathbb{B}} |d\gamma|^s \right)^{\frac{1}{s}} \preccurlyeq \|\alpha\| \left[\left(\fint_{\mathbb{B}} |Df|^{\frac{n^2}{n+1}} \right)^{\frac{n+1}{n^2}} + \|Df\|_n \right]^l \tag{7.31}$$

Note that $\left(\fint_{\mathbb{B}} |Df|^{\frac{n^2}{n+1}} \right)^{\frac{n+1}{n^2}} \leqslant \mathbf{M}_p(Df)$, since $p = \frac{2n^2}{2n+1} > \frac{n^2}{n+1}$. This combined with (7.31), (7.27) and (7.29) results in the following estimate

$$\left| (d\gamma \wedge f^\sharp \beta)_t \right| \preccurlyeq \|\alpha\|_\infty \|\beta\|_\infty \left[\|Df\|_n + \mathbf{M}_p(Df) \right]^n \tag{7.32}$$

By virtue of the previously established inequality for $\mathcal{M}(\chi \wedge f^\sharp \beta)$, we obtain

(7.33) $$\left| (f^\sharp \omega)_t \right| \preccurlyeq \|\alpha\|_\infty \|\beta\|_\infty \Big[\|Df\|_n + \mathbf{M}_p(Df) \Big]^n$$

for $0 < t \leqslant R_\mathbb{X}$. In summary, if $\omega = \alpha \wedge \beta$ is a Cartan form, then

(7.34) $$\mathcal{M}(f^\sharp \omega) \preccurlyeq \|\alpha\|_\infty \|\beta\|_\infty \Big[\|Df\|_n + \mathbf{M}_p(Df) \Big]^n$$

Hence, Remark 7.4 is straightforward by the maximal theorem.

7.2.2. Step 2.-The case of closed forms. Now, we prove Theorem 7.3 for closed forms. As pointed out in Remark 7.5 we need only establish the following inequality

(7.35) $$\left| \int_\mathbb{X} K_t(a,x) \mathcal{J}_\omega(x,f)\, dx \right| \preccurlyeq C(\omega, g) \Big[\mathbf{M}_p(|Df|)(a) + \|Df\|_n \Big]^n$$

for all $t > 0$, $a \in \mathbb{X}$ and $f \in \mathscr{B}(g, \epsilon)$. Here $\mathcal{J}_\omega(\cdot, f)$ stands for the pointwise pullback $f^\sharp \omega$.

We shall work with small balls $\mathbb{B}(a, t) \subset \mathbb{X}$; say with radii $t \leqslant R = R(g)$. Let us begin with a clear list of the conditions on R and ϵ needed in the sequel. First condition is:

(7.36) $$R \leqslant R_\mathbb{X}.$$

Another restriction on ϵ and R results from the following lemma

LEMMA 7.7. [OSCILLATION LEMMA] *Let $h \in \mathscr{C}^\infty(\mathbb{X}, \mathbb{Y})$ and let $\mathbb{B}(a, R)$ be a legitimate ball in $\mathbb{X}$. Then for every $0 < t < 2t \leqslant R$ there exists $r \in (t, 2t)$ such that*

(7.37) $$\operatorname*{osc}_{\partial \mathbb{B}(a,r)} h \preccurlyeq r \left(\fint_{\mathbb{B}(a,2r)} |Dh|^p \right)^{\frac{1}{p}}$$

where $n - 1 < p = \frac{2n^2}{n+1} < n$. By Hölder's inequality, there is a constant $C(n\mathbb{X})$ such that

(7.38) $$\left(\operatorname*{osc}_{\partial \mathbb{B}(a,r)} h \right)^n \leqslant C(n, \mathbb{X})^n \int_{\mathbb{B}(a,R)} |Dh|^n$$

We want these oscillations to be smaller than $R_\mathbb{Y}$. For this reason we must confine ourselves to $R < R_\mathbb{X}$ and ϵ small enough so that

(7.39) $$C(n, \mathbb{X})^n \int_{\mathbb{B}(a,R)} |Dg(x)|^n \, dx \leqslant \left(\frac{1}{2} R_\mathbb{Y} \right)^n$$

for all $a \in \mathbb{X}$ and, in addition, we assume that

(7.40) $$2 C(n, \mathbb{X}) \epsilon < R_\mathbb{Y}$$

Next we wish that the integrals $\int_{\mathbb{B}(a,R)} |Df|^n$, with $a \in \mathbb{X}$ and $f \in \mathscr{B}(g, \epsilon)$ be sufficiently small. To determine ϵ we first note the inequality

(7.41) $$\begin{aligned} \left(\fint_{\mathbb{B}(a,R)} |Df|^n \, dx \right)^{1/n} &\leqslant \left(\fint_{\mathbb{B}(a,R)} |Dg|^n \, dx \right)^{1/n} \\ &\quad + \left(\fint_{\mathbb{B}(a,R)} |Df - Dg|^n \, dx \right)^{1/n} \\ &\leqslant \left(\fint_{\mathbb{B}(a,R)} |Dg|^n \, dx \right)^{1/n} + \epsilon \end{aligned}$$

for each $a \in \mathbb{X}$. A theorem of B. White (Theorem 2. p. 135 in [**56**]) states that for each $g \in \mathscr{W}^{1,n}(\mathbb{X}, \mathbb{Y})$ there exists $\rho > 0$ such that if $f_1, f_2 \in \mathscr{B}(g, 2\rho)$ are smooth for $i = 1, 2$, then f_1 and f_2 are homotopic. The requirement

$$(7.42) \qquad C(\mathbb{Y})\left[\left(\fint_{\mathbb{B}(a,R)} |Dg|^n\, dx\right)^{1/n} + \epsilon\right] \leqslant \rho$$

where the constant $C(\mathbb{Y})$ will be determined later, see (7.53). This is the last condition on R and ϵ.

The above conditions at (7.36)–(7.42) determine the numbers $R = R(g) > 0$ and $\epsilon = \epsilon(g) > 0$. The remaining estimates in this section will be explicit given $R = R(g)$ and $\epsilon = \epsilon(g)$. Returning to the inequality (7.35) let us temporarily fix both $f \in \mathscr{B}(g, \epsilon)$ and the parameter $0 < t < R(g)$. It involves no loss of generality in assuming that $f \in \mathscr{C}^\infty(\mathbb{X}, \mathbb{Y})$, simply because $\mathscr{C}^\infty(\mathbb{X}, \mathbb{Y})$ is dense in $\mathscr{W}^{1,n}(\mathbb{X}, \mathbb{Y})$. It remains to prove the inequality

$$(7.43) \qquad \left|\int_{\mathbb{X}} K_t(a, x) \mathcal{J}_\omega(x, f)\right| \preccurlyeq \left[\mathbf{M}_p(|Df|)(a)\right]^n$$

where the implied constant depends on R which we have already determined for the given function g.

We fix finite covering of $\mathbb{Y}$ by legitimate balls of radius $T \stackrel{\text{def}}{=\!=} R_\mathbb{Y}$. Using the oscillation inequality at (7.37) we find a radius $r \in (t, 2t)$ such that

$$(7.44) \qquad \operatorname*{osc}_{\partial \mathbb{B}(a,r)} f \leqslant R_\mathbb{Y} = T$$

which is immediate from (7.38). We look at the image of $f : \partial \mathbb{B}(a, r) \to \mathbb{Y}$. It intersects some legitimate ball $\mathbb{B}(b, T) \subset \mathbb{Y}$ from the above mentioned finite cover of $\mathbb{Y}$. Then by (7.44),

$$(7.45) \qquad f(\partial \mathbb{B}(a, r)) \subset \mathbb{B}(b, 2T).$$

Recall that there is a coordinate chart $(\Omega, \kappa) \in \mathcal{A}$ in which $\mathbb{B}(b, 4T) \subset \Omega$ and $\kappa : \Omega \xrightarrow{\text{onto}} \mathbb{R}^m$. We may assume that $\kappa(b) = 0$. Consider a cut-off function $\eta \in \mathscr{C}^\infty(\mathbb{Y})$ with support in $\mathbb{B}(b, 4T)$ and equal to 1 on a neighborhood of $\mathbb{B}(b, 3T)$. The form $(\kappa^{-1})^\sharp \omega$ is closed in $\mathbb{R}^m$ and thus there exists a form $\gamma \in \mathscr{C}^\infty(\wedge^{n-1}\mathbb{R}^m)$ such that $d\gamma = (\kappa^{-1})^\sharp \omega$. We have the identity $\omega = \kappa^\sharp(d\gamma) = d(\kappa^\sharp \gamma)$, hence

$$(7.46) \qquad \eta\omega = d(\eta \cdot \kappa^\sharp \gamma) - d\eta \wedge \kappa^\sharp \gamma$$

Consider the form

$$\tilde{\omega} = d(\eta \cdot \kappa^\sharp \gamma)$$

Then $\tilde{\omega}$ is exact and coincides with ω on $\mathbb{B}(b, 3T)$. Since the legitimate ball is selected from the given finite family, all quantities related to η and κ depend only on $\mathbb{Y}$, and all quantities related to $\tilde{\omega}$ depend only on $\mathbb{Y}$ and ω. Define

$$\tilde{f}(x) = \begin{cases} \kappa^{-1}\Big(\kappa(z) + \eta(f(x))\big(\kappa(f(x)) - \kappa(z)\big)\Big) & f(x) \in \Omega \\ b & f(x) \notin \Omega \end{cases}$$

where z is a point of $\mathbb{Y}$ which is nearest to the mean value of f. Then

$$\mathcal{J}_\omega(x, \tilde{f}) = \mathcal{J}_{\tilde{\omega}}(x, f)$$

7.2. THE $\mathscr{H}^1$-THEOREM

and by (7.34) we have

$$
\begin{aligned}
\left|\int_{\mathbb{X}} K_t(a,x)\mathcal{J}_\omega(x,\tilde{f})\,dx\right| &\leqslant C(\omega)\Big[\,\|D\tilde{f}\,\|_n + \mathbf{M}_p(D\tilde{f})(a)\Big]^n \\
&\leqslant C(\omega)\Big[\,\|Df\|_n + \|f-z\|_n + \mathbf{M}_p(Df)(a)\Big]^n \\
&\leqslant C(\omega)\Big[\,\|Df\|_n + \mathbf{M}_p(Df)(a)\Big]^n
\end{aligned}
\tag{7.47}
$$

where in the last step we have used a version of the Poincaré inequality.

It remains to prove the estimate

$$\left|\int_{\mathbb{X}} K_t(a,x)\Big(\mathcal{J}_\omega(x,\tilde{f}) - \mathcal{J}_\omega(x,f)\Big)\,dx\right| \leq C(\omega,g)\Big[\mathbf{M}_p(Df)(a)\Big]^n \tag{7.48}$$

Let us look closely at the set

$$\mathbb{E} \stackrel{\text{def}}{=} \mathbb{B}(a,r) \cap f^{-1}\Big(\mathbb{Y}\setminus \mathbb{B}(b,3T)\Big) \supset \mathbb{B}(a,r) \cap \Big\{\mathcal{J}_\omega(x,\tilde{f}) \neq \mathcal{J}_\omega(x,f)\Big\}$$

We first notice that $f(\partial\mathbb{B}(a,r))$ lies in $\mathbb{B}(b,2T)$ by (7.45). Thus $\mathbb{E}$ is compact subset of the ball $\mathbb{B}(a,r)$. The function

$$u = \frac{|f-b|}{T} - 2$$

is negative on $\partial\mathbb{B}(a,r)$ and assumes values ≥ 1 on the set $\mathbb{E}$. By Corollary 7.2 we can split $\mathbb{E}$ into mutually disjoint compact sets $\mathbb{E}_1,...,\mathbb{E}_k$ such that

$$
\begin{aligned}
\sum_{i=1}^k \text{diam}\,\mathbb{E}_i &\leqslant C_{\mathbb{X}}\, r^{n+1}\left(\fint_{\mathbb{B}(a,r)} |\nabla u|^p\right)^{\frac{n}{p}} \\
&\preccurlyeq t^{n+1}\Big[\mathbf{M}_p(|Df|)(a)\Big]^n
\end{aligned}
\tag{7.49}
$$

We accordingly split the integral at (7.48) as:

$$
\begin{aligned}
\int_{\mathbb{B}(a,t)} &K_t(a,x)\Big(\mathcal{J}_\omega(x,\tilde{f}) - \mathcal{J}_\omega(x,f))\Big)\,dx \\
&= \sum_{i=1}^k \int_{\mathbb{E}_i} K_t(a,x)\Big(\mathcal{J}_\omega(x,\tilde{f}) - \mathcal{J}_\omega(x,f))\Big)\,dx
\end{aligned}
\tag{7.50}
$$

An important point to make here is that

$$\int_{\mathbb{E}_i} \mathcal{J}_\omega(x,\tilde{f})\,dx = \int_{\mathbb{E}_i} \mathcal{J}_\omega(x,f)\,dx, \quad \text{for } i=1,2,...,k \tag{7.51}$$

To see this we consider the following functions

$$f_i = \begin{cases} \tilde{f} & \text{on } \mathbb{E}_i \\ f & \text{on } \mathbb{X}\setminus\mathbb{E}_i \end{cases} \tag{7.52}$$

Then f_i are smooth. Using (7.42) we obtain

$$\|f - f_i\|_{1,n} \leqslant C(\mathbb{Y})\left(\int_{B(a,r)} |Df|^n\right)^{1/n} \tag{7.53}$$

Thus both functions f and f_i belong to $\mathscr{B}(g, 2\rho)$. This, by the definition of ρ, implies that f and f_i are homotopic. Hence
$$\int_{\mathbb{X}} \mathcal{J}_\omega(x, f_i)\, dx = \int_{\mathbb{X}} \mathcal{J}_\omega(x, f)\, dx$$
which proves the claim (7.51). Similarly we obtain the estimate

(7.54) $$\int_{\mathbb{E}_i} |\mathcal{J}_\omega(x, \tilde{f}) - \mathcal{J}_\omega(x, f)|\, dx \preccurlyeq \int_{B(a,r)} |Df|^n \preccurlyeq \rho$$

To make use of the formulas (7.49) we pick up some points $x_i \in \mathbb{E}_i$ and express the integral on the left hand side of (7.48) as

(7.55) $$\int_{\mathbb{B}(a,t)} K_t(a, x)\bigl(\mathcal{J}_\omega(x, \tilde{f}) - \mathcal{J}_\omega(x, f)\bigr)\, dx$$
$$= \sum_{i=1}^{k} \int_{\mathbb{E}_i} \bigl[K_t(a, x) - K_t(a, x_i)\bigr]\bigl(\mathcal{J}_\omega(x, \tilde{f}) - \mathcal{J}_\omega(x, f)\bigr)\, dx$$

Next, we use the following inequalities for the mollifiers

(7.56) $$|K_t(a, x) - K_t(a, x_i)| \preccurlyeq \frac{|x - x_i|}{t^{n+1}} \leqslant \frac{\operatorname{diam} \mathbb{E}_i}{t^{n+1}}$$

for all $x \in \mathbb{E}_i$. They follow routinely from (2.45). Finally, by (7.49) and (7.54) we conclude with the desired estimate

(7.57) $$\left| \int_{\mathbb{B}(a,t)} K_t(a,x)\bigl(\mathcal{J}_\omega(x, \tilde{f}) - \mathcal{J}_\omega(x, f)\bigr)\, dx \right|$$
$$\preccurlyeq \sum_{i=1}^{k} \frac{\operatorname{diam} \mathbb{E}_i}{t^{n+1}} \int_{\mathbb{E}_i} |\mathcal{J}_\omega(x, \tilde{f}) - \mathcal{J}_\omega(x, f)|\, dx$$
$$\preccurlyeq \bigl[\mathbf{M}_p(Df)(a)\bigr]^n$$

completing the proof of (7.48) and thus of Theorem 7.3.

CHAPTER 8

Degree Theory

$\mathscr{L}^1$-estimates of the Jacobian and related wedge products lead to an analytic degree theory of weakly differentiable mappings. Readers interested in this topic will find it profitable to consult Brezis and Nirenberg [6], [7] and also [4], [5], [11], [22], [14]. Analytic approach to the degree of smooth mappings begins with a choice of a closed form $\omega \in \mathscr{C}^\infty(\wedge^n \mathbb{Y})$. This form must have non-vanishing integral, which may not be possible within the class of Cartan forms;

$$(8.1) \qquad \omega = \sum_{i=1}^{K} \alpha_i \wedge \beta_i, \quad d\alpha_i = d\beta_i = 0, \quad \int_{\mathbb{Y}} \omega \neq 0,$$

Unluckily, such is the case $\mathbb{Y} = \mathbb{S}^n$. On the other hand we need Cartan forms in order to employ Theorems 6.5, 6.6 and 6.7. More generally, if $\mathscr{H}^l(\mathbb{Y}) = 0$ for all $1 \leqslant l < n$, Cartan's forms are exact and, therefore, have integral zero. This being so, we must assume that $\mathscr{H}^l(\mathbb{Y}) \neq 0$, for some $1 \leqslant l < n$.

8.1. Definition of the degree via weak integrals

There are several approaches to the degree of Sobolev mappings that are each of considerable interest. We shall give first the most general one.

DEFINITION 8.1. Let $\dim \mathbb{X} = \dim \mathbb{Y} = n$. The notation and hypothesis being as in Theorem 6.5, we define the degree of $f : \mathbb{X} \to \mathbb{Y}$ by the rule

$$(8.2) \qquad \deg(f; \mathbb{X}, \mathbb{Y}) = \lim_{j \to \infty} \int_{\mathbb{X}_j} \mathcal{J}_\omega(x, f)\, dx = (f^\sharp \omega)[1]$$

where $\omega \in \mathscr{C}^\infty(\wedge^n \mathbb{Y})$ has integral 1 over $\mathbb{Y}$.

Absence of ω in the notation for the degree is justified by Theorem 8.2 below.

From differential topology we know that the notion of the topological degree of a mapping $f : \mathbb{X} \to \mathbb{Y}$ of class $\mathscr{C}^1(\mathbb{X}, \mathbb{Y})$ coincides with the integral of $\mathcal{J}_\omega(x, f)$ and, therefore, is an integer. Basic characteristics of $\deg(f; \mathbb{X}, \mathbb{Y})$; that justify the name degree, are listed in the following theorem.

THEOREM 8.2. *With the reference to Definition 8.1, we have*
  (i) *Different choices of the Cartan forms with integral 1 yield the same limit at (8.2).*
  (ii) *If smooth mappings $f_k : \mathbb{X} \to \mathbb{Y}$, converge to f in $\mathscr{W}^{1,s}(\mathbb{X}, \mathbb{Y})$, with the critical exponent $s = \frac{n^2}{n+1}$, then*

$$\deg(f; \mathbb{X}, \mathbb{Y}) = \lim_{k \to \infty} \deg(f_k; \mathbb{X}, \mathbb{Y})$$

  Moreover, such a smooth approximation of f always exists.
  (iii) *The degree is an integer*

(iv) *If the Jacobian $\mathcal{J}_\omega(x,f)$ is non-negative, then it is integrable and we have*
$$\deg(f; \mathbb{X}, \mathbb{Y}) = \int_\mathbb{X} \mathcal{J}_\omega(x,f)\,dx$$

(v) *If $\deg(f; \mathbb{X}, \mathbb{Y}) \neq 0$, then the image of any set of full measure is dense in $\mathbb{Y}$*

PROOF. The proof is simply an adaptation of ideas of analytic degree theory of smooth mappings. To prove statement (i) we fix two Cartan forms $\omega, \theta \in \mathscr{C}^\infty(\wedge^n \mathbb{Y})$ whose integral is equal to 1. Thus

(8.3) $$\lim_{j \to \infty} \int_{\mathbb{X}_j} \mathcal{J}_\omega(x,f)\,dx = (f^\sharp \omega)[1]$$

and

(8.4) $$\lim_{j \to \infty} \int_{\mathbb{X}_j} \mathcal{J}_\theta(x,f)\,dx = (f^\sharp \theta)[1]$$

by Theorem 6.5. Now the problem reduces to showing that $(f^\sharp \omega)[1] = (f^\sharp \theta)[1]$. Thanks to Theorem 1.2, we can approximate f by smooth mappings in the metric topology of $\mathscr{W}^{1,s}(\mathbb{X}, \mathbb{Y})$. Since the nonlinear functionals $f \to (f^\sharp \omega)[1]$ and $f \to (f^\sharp \theta)[1]$ are continuous in $\mathscr{W}^{1,s}(\mathbb{X}, \mathbb{Y})$, we are further reduced to showing that $(f^\sharp \omega - f^\sharp \theta)[1] = 0$, whenever $f \in \mathscr{C}^\infty(\mathbb{X}, \mathbb{Y})$. To this end we observe that the differential form $\omega - \theta \in \mathscr{C}^\infty(\wedge^n \mathbb{Y})$ is exact; that is $\omega - \theta = d\alpha$ for some $\alpha \in \mathscr{C}^\infty(\wedge^{n-1} \mathbb{Y})$. This is because the integral of $\omega - \theta$ over $\mathbb{Y}$ vanishes and $\mathcal{H}^n(\mathbb{Y}) \simeq \mathbb{R}$. The rest is folklore, $f^\sharp(d\alpha) = d(f^\sharp \alpha)$ and by Stokes' theorem

(8.5) $$(f^\sharp \omega - f^\sharp \theta)[1] = \int_\mathbb{X} d(f^\sharp \alpha) = 0$$

The property (ii) is immediate from Theorem 1.2. Then to see (iii), we need only recall that the degree of a smooth mapping is an integer. Also (iv) follows readily from Theorem 6.6.

As for the statement (v), consider a set $\mathbb{X}'$ of full measure in $\mathbb{X}$. Let us assume, to the contrary, that $f : \mathbb{X}' \to \mathbb{Y}$ omits an open set $\mathbb{V} \subset \mathbb{Y}$. Fix a Cartan form $\omega \in \mathscr{C}_0^\infty(\wedge^n \mathbb{V})$ whose integral over $\mathbb{Y}$ equals 1; for instance, $\omega = \lambda(y)\,dy$ with $\lambda \in \mathscr{C}_0^\infty(\mathbb{V})$. Thus $\mathcal{J}_\omega(x,f) = 0$ almost everywhere in $\mathbb{X}'$, hence in $\mathbb{X}$ as well. Being so, the limit at (8.2) is equal to zero, contradicting the assumption that $\deg(f; \mathbb{X}, \mathbb{Y}) \neq 0$.

8.2. Weak integrals

Our next objective is to investigate properties of the degree function $f \to \deg(f; \mathbb{X}, \mathbb{Y})$ defined on mappings $f \in \mathscr{W}^{1,s}(\mathbb{X}, \mathbb{Y})$, $s = \frac{n^2}{n+1}$, such that

(8.6) $$\liminf_{t \to \infty} t^{n-s} \int_{|Df| > t} |Df|^s = 0$$

We assume here that $\mathbb{Y}$ has nontrivial l-cohomology for some $1 \leqslant l < n = \dim \mathbb{Y} = \dim \mathbb{X}$. As a preliminary step we consider a nonlinear functional $\mathcal{J}_\omega : \mathscr{W}^{1,s}(\mathbb{X}, \mathbb{Y}) \to \mathbb{R}$, defined by

(8.7) $$\mathcal{J}_\omega[f] = (f^\sharp \omega)[1]$$

where $\omega \in \mathscr{C}^\infty(\wedge^n \mathbb{Y})$ is fixed. We call it *weak integral* of the Jacobian.

8.2. WEAK INTEGRALS

8.2.1. Continuity of the weak integral. Surprisingly, $\mathcal{J}_\omega$ is continuous even in the metric topology of $\mathscr{W}^{1,n-1}(\mathbb{X}, \mathbb{Y})$.

LEMMA 8.3. *Suppose* $f_\nu \in \mathscr{W}^{1,s}(\mathbb{X}, \mathbb{Y})$ *converge to*
$$f \in \mathscr{W}^{1,s}(\mathbb{X}, \mathbb{Y}) \subset \mathscr{W}^{1,n-1}(\mathbb{X}, \mathbb{Y})$$
in the metric of $\mathscr{W}^{1,n-1}(\mathbb{X}, \mathbb{Y})$. *Then*
$$(8.8) \qquad \lim_{\nu \to \infty} \mathcal{J}_\omega[f_\nu] = \mathcal{J}_\omega[f]$$

8.2.2. $\mathscr{L}^1$-Estimate of the weak integral. Before jumping to the proof of Lemma 8.3 let us state another surprising result, which will be the key ingredient.

LEMMA 8.4. *Given* $\Phi \in \mathscr{L}^p(\wedge^l \mathbb{X}) \cap \ker d$ *and* $\Psi \in \mathscr{L}^r(\wedge^k \mathbb{X}) \cap \ker d$, $1 \leqslant k, l < n$, $k + l = n$, *where* $1 \leqslant p, r < \infty$ *are Sobolev conjugate exponents. We have*
$$(8.9) \qquad |(\Phi \wedge \Psi)[1]| \preccurlyeq \|\Phi\|_{\mathscr{L}^1(\mathbb{X})} \|\Psi\|_{\mathscr{L}^1(\mathbb{X})}$$

This estimate is not always true if we replace 1 by arbitrary test function $\eta \in \mathscr{C}^\infty(\mathbb{X})$.

PROOF. By the definition of the distributional wedge product, given at (6.15), it will be enough to prove (8.9) for smooth forms. In this case, we have
$$(8.10) \qquad (\Phi \wedge \Psi)[1] = \int_{\mathbb{X}} \Phi \wedge \Psi$$

If one of the factors Φ or Ψ is exact then so is their wedge product. In this case the integral vanishes, so there is nothing to estimate. But this is not always the case. Fortunately, closed forms are exact modulo harmonic fields, which we consider as harmless terms. Precisely, we proceed as follows:
$$(8.11) \qquad \Phi = d\varphi + h, \quad h \in \mathcal{H}^l(\mathbb{X}) \text{ and } \varphi \in \mathscr{C}^\infty(\wedge^{l-1}\mathbb{X})$$

Although we may not have good estimates of $d\varphi$ in terms of Φ, we do have, however, good estimates of the harmonic component. Luckily, $d\varphi$ disappears after we integrate at (8.10):
$$(8.12) \qquad \int_{\mathbb{X}} \Phi \wedge \Psi = \int_{\mathbb{X}} h \wedge \Psi$$
by Stokes' theorem. Hence
$$(8.13) \qquad |(\Phi \wedge \Psi)[1]| \leqslant \|h\|_{\mathscr{L}^\infty(\mathbb{X})} \|\Psi\|_{\mathscr{L}^1(\mathbb{X})}$$

The rest of the proof relies on the regularity properties of the harmonic fields, see inequality (2.15). Accordingly,
$$(8.14) \qquad \|h\|_\infty \preccurlyeq \sup_{t>0} t |\{x;\ |h(x)| > t\}| \preccurlyeq \|\Phi\|_1$$
as desired.

8.2.3. Proof of Lemma 8.3. As we have already observed in Lemma 2.13 every $\omega \in \mathscr{C}^\infty(\wedge^n \mathbb{Y})$ is a Cartan form, say
$$(8.15) \qquad \omega = \sum_{i=1}^K \alpha_i \wedge \beta_i$$

Hence $(f_\nu^\sharp \omega)[1] = \sum_{i=1}^K (\Phi_i^\nu \wedge \Psi_i^\nu)[1]$, where both $\Phi_i^\nu = f_\nu^\sharp \alpha_i$ and $\Psi_i^\nu = f_\nu^\sharp \beta_i$ are closed forms of degree $1 \leqslant l_i < n$ and $1 \leqslant k_i < n$, respectively. Similarly, $(f^\sharp \omega)[1] = \sum_{i=1}^K (\Phi_i \wedge \Psi_i)[1]$. First observe the point-wise inequalities $|\Phi_i^\nu| \preccurlyeq |Df_\nu|^{l_i}$ and

$|\Psi_i^\nu| \preccurlyeq |Df_\nu|^{k_i}$, $l_i + k_i = n$. Thus $\Phi_i^\nu \in \mathscr{L}^{p_i}(\wedge^{l_i}\mathbb{X})$ and $\Psi_i^\nu \in \mathscr{L}^{r_i}(\wedge^{k_i}\mathbb{X})$, with a Sobolev conjugate pair of exponents $p_i = \frac{s}{l_i}$ and $r_i = \frac{s}{k_i}$, $\frac{1}{p_i} + \frac{1}{r_i} = \frac{n}{s} = 1 + \frac{1}{n}$. We need to show that

$$(8.16) \qquad \lim_{\nu \to \infty} (\Phi_i^\nu \wedge \Psi_i^\nu)[1] = (\Phi_i \wedge \Psi_i)[1]$$

for every $i = 1, 2, \ldots, K$. Using telescoping decomposition, this reduces to two equations:

$$(8.17) \qquad \lim_{\nu \to \infty} \left[(\Phi_i^\nu - \Phi_i) \wedge \Psi_i^\nu \right][1] = 0$$

and

$$(8.18) \qquad \lim_{\nu \to \infty} \left[\Phi_i \wedge (\Psi_i^\nu - \Psi_i) \right][1] = 0$$

We will only demonstrate the proof of (8.17); the other being similar is omitted. By Lemma 8.4, we have

$$(8.19) \qquad \begin{aligned} \left| \left[(\Phi_i^\nu - \Phi_i) \wedge \Psi_i^\nu \right][1] \right| &\preccurlyeq \|\Phi_i^\nu - \Phi_i\|_{\mathscr{L}^1(\mathbb{X})} \|\Psi_i^\nu\|_{\mathscr{L}^1(\mathbb{X})} \\ &\preccurlyeq \|f_\nu^\sharp \alpha_i - f_\nu^\sharp \alpha\|_{\mathscr{L}^1(\mathbb{X})} \|Df_\nu\|_{\mathscr{L}^{k_i}(\mathbb{X})}^{k_i} \end{aligned}$$

Since $k_i \leqslant n-1$ the last factor is bounded by $\|Df_\nu\|_{n-1}^{k_i}$. Next observe the pointwise inequality

$$(8.20) \qquad |f_\nu^\sharp \alpha_i - f_\nu^\sharp \alpha| \preccurlyeq |Df_\nu - Df|(|Df_\nu| + |Df|)^{l_i - 1} + |f_\nu - f||Df|^{l_i}$$

This can be easily verified using local coordinates. The $\mathscr{L}^1$-norm of the first term in the right hand side of (8.20) is controlled by

$$\|Df_\nu - Df\|_{n-1} \Big(\|Df_\nu\|_{n-1} + \|Df\|_{n-1} \Big)^{l_i - 1}$$

Simple application of Hölder's inequality shows that integral of the second term is bounded by

$$(8.21) \qquad \int_\mathbb{X} |f_\nu - f||Df|^{l_i} \preccurlyeq \left(\int_\mathbb{X} |f_\nu - f|^{n-1} |Df|^{n-1} \right)^{\frac{1}{n-1}} \left(\int_\mathbb{X} |Df|^{n-1} \right)^{\frac{l_i - 1}{n-1}}$$

We conclude with the following inequality

$$(8.22) \qquad \left| (f_\nu^\sharp \omega - f^\sharp \omega)[1] \right|$$
$$\preccurlyeq \left(\| |f_\nu - f||Df| \|_{n-1} + \|Df_\nu - Df\|_{n-1} \right) \left(\int_\mathbb{X} |Df_\nu|^{n-1} + |Df|^{n-1} \right)$$

Finally, let ν go to infinity. The integral stays bounded and the term $\|Df_\nu - Df\|_{n-1}$ goes to zero. Also $(f_\nu - f)|Df|$ goes to zero in $\mathscr{L}^{n-1}$ by the Lebesgue Convergence Theorem. Hence $\lim_{\nu \to \infty} (f_\nu^\sharp \omega)[1] = (f^\sharp \omega)[1]$, as desired.

8.3. Stability of the degree

Next we are concerned with the fundamental question of the degree theory; how close should the mappings $f, g \in \mathscr{C}^\infty(\mathbb{X}, \mathbb{Y})$ be in order to ensure that they have the same degree. We shall measure the distance using the metric of the Sobolev space $\mathscr{W}^{1,q}(\mathbb{X}, \mathbb{Y})$ with $q > n-1$. We also assume, as always, that the l-cohomology of the target space is nontrivial for some $1 \leqslant l < n$.

THEOREM 8.5. *Given $M > 0$ and $q > n - 1$, there exists $\epsilon = \epsilon(\mathbb{X}, \mathbb{Y})$ such that if two mappings $f, g \in \mathscr{C}^\infty(\mathbb{X}, \mathbb{Y})$ satisfy*

(8.23) $$\|f\|_{\mathscr{W}^{1,q}} + \|g\|_{\mathscr{W}^{1,q}} \leqslant M \quad \text{and} \quad \|f - g\|_{\mathscr{W}^{1,q}} \leqslant \epsilon$$

Then $\deg(f; \mathbb{X}, \mathbb{Y}) = \deg(g; \mathbb{X}, \mathbb{Y})$.

PROOF. The reader may carefully reexamine the proof of (8.22) to observe that we have actually proven the following estimate

$$\left| (f^\sharp \omega - g^\sharp \omega)[1] \right|$$
$$\preccurlyeq \left(\| |f - g| |Df| \|_{n-1} + \|Df - Dg\|_{n-1} \right) \left(\|Df\|_{n-1}^{n-1} + \|Dg\|_{n-1}^{n-1} \right)$$

whenever $f, g \in \mathscr{C}^\infty(\mathbb{X}, \mathbb{Y})$ and $q > n - 1$. Since the target space is bounded, it folows

(8.24) $$\left| (f^\sharp \omega - g^\sharp \omega)[1] \right| \preccurlyeq \|f - g\|_{\mathscr{W}^{1,q}} \left(\|f\|_{\mathscr{W}^{1,q}} + \|g\|_{\mathscr{W}^{1,q}} \right)^{n-1}$$

This proves Theorem 8.5.

REMARK 8.6. Theorem 8.5 also holds for mappings $f, g \in \mathscr{W}^{1,s}(\mathbb{X}, \mathbb{Y})$, provided they both satisfy condition (6.20). This is because we could approximate them by smooth mappings.

8.4. The degree in Orlicz and grand Sobolev spaces

Finally, our discussion is narrowed to Orlicz-Sobolev and to grand-Sobolev classes of mappings $f : \mathbb{X} \to \mathbb{Y}$, $\dim \mathbb{X} = \dim \mathbb{Y} = n$, where $\mathcal{H}^l(\mathbb{Y}) \neq 0$ for some $1 \leqslant l < n$. Recall that these classes hold smooth approximation property, by Theorems 5.2 and 1.3.

Let P satisfy the hypothesis of Theorem 5.2. As a corollary of Theorem 8.5, we conclude:

THEOREM 8.7. *The degree function*

(8.25) $$\deg : \mathscr{W}^{1,P}(\mathbb{X}, \mathbb{Y}) \to \{..., -2, -1, 0, 1, 2, ...\}$$

is uniformly continuous on every bounded subclass of $\mathscr{W}^{1,P}(\mathbb{X}, \mathbb{Y})$.

Speaking of the category of grand Sobolev spaces, let us recall that

(8.26) $$\lim_{\epsilon \to 0} \epsilon \int_\mathbb{X} |Df(x)|^{n-\epsilon} dx = 0$$

whenever $f \in V\mathscr{W}^{1,n}(\mathbb{X}, \mathbb{Y})$. For such mappings we have yet another interesting integral formula for the degree

(8.27) $$\deg(f; \mathbb{X}, \mathbb{Y}) = \lim_{\epsilon \to 0} \int_\mathbb{X} \frac{\mathcal{J}_\omega(x, f) \, dx}{|\mathcal{J}_\omega(x, f)|^\epsilon}$$

simply because this limit coincides with $(f^\sharp \omega)[1]$, see [14], [34]. The advantage of this latter formula is that we neither approximate f by smooth mappings nor approximate $\mathbb{X}$ by the sets $\mathbb{X}_j$. This formula might be extremely useful in numerical treatment of the degree theory. Indeed, as $\deg(f; \mathbb{X}, \mathbb{Y})$ is an integer, we only need to compute the limit at (8.27) with sufficient accuracy to ensure that the error is less than $\frac{1}{2}$. Explicit estimates of the error in terms of ϵ are also available.

One particular Orlicz-Sobolev subspace of $V\mathscr{W}^{1,n}(\mathbb{X},\mathbb{Y})$ deserves mentioning here. This is the class of weakly differentiable mapping $f:\mathbb{X}\to\mathbb{Y}$ whose differential lies in the Zygmund class $\mathscr{L}^n\log^{-1}\mathscr{L}(\mathbb{X})$; that is

$$(8.28)\qquad \int_\mathbb{X} \frac{|Df(x)|^n\,dx}{\log(e+|Df(x)|)} < \infty$$

CHAPTER 9

Mappings of Finite Distortion

Recently there have been considerable advances made in the study of mappings of finite distortion between the domains in $\mathbb{R}^n$. The reader interested in these developments is referred to [**30**], [**38**], [**39**], [**31**] and the recent monograph [**32**]. What we want to point out here is the extent to which those results are true in the Riemannian manifold setting.

DEFINITION 9.1. Let $\dim \mathbb{X} = \dim \mathbb{Y} = n$. A Sobolev mapping $f : \mathbb{X} \to \mathbb{Y}$ is said to have finite distortion if
  (i) The Jacobian determinant $\mathcal{J}(x, f) \, dx = f^\sharp(dy)$ is integrable
  (ii) There is a measurable function $K = K(x) \geqslant 1$, finite almost everywhere, such that f satisfies the distortion inequality

$$(9.1) \qquad |Df(x)|^n \leqslant K(x) \, \mathcal{J}(x, f) \quad \text{for almost every } x \in \mathbb{X}$$

We emphasize that in many natural situations the condition (i) is automatic. Such is the case when f is a local homeomorphism. More generally, $\mathcal{J}(x, f) \in \mathscr{L}^1(\mathbb{X})$ if the cardinality of the set $\{x \in \mathbb{X}; \, f(x) = y\}$ is an integrable function in $y \in \mathbb{Y}$. Foundational analysis of mappings of finite distortion relies on integration of the Jacobian. In order to fully benefit from the estimates and the degree formulas we must stay close to the natural Sobolev class $\mathscr{W}^{1,n}(\mathbb{X}, \mathbb{Y})$. Thanks to $\mathscr{L}^1$-estimates in Section 6 we may consider unbounded distortion $K = K(x)$. It turns out that the following integral condition on K has interesting implications

$$(9.2) \qquad \int_\mathbb{X} e^{\Phi(K(x))} \, dx < \infty, \quad \text{where} \quad \int_1^\infty \frac{\Phi(t)}{t^2} \, dt = \infty$$

This implies, via the distortion inequality, that $f \in \mathscr{W}^{1,P}(\mathbb{X}, \mathbb{Y})$, where P satisfies the hypotheses of Theorem 8.5. To be precise, we should mention here that one also needs $\Phi(t) \succcurlyeq \log t$. This additional condition plays rather technical role, since in practice $\Phi(t)$ behaves more or less like the linear function. For instance, $\Phi(t) = \lambda t$ or $\Phi(t) = \lambda t \log^{-1}(e + t)$, $\lambda > 0$. As a consequence of our investigation of the pullback of Sobolev mappings we are able to carry out this program on manifolds.

THEOREM 9.2. Let $f : \mathbb{X} \to \mathbb{Y}$ be a non-constant mapping of finite distortion $K = K(x)$ satisfying (9.2). Then
  - f is continuous, open, and discrete.
  - The measure of $\mathbb{E} \subset \mathbb{X}$ is zero if and only if $f(\mathbb{E}) \subset \mathbb{Y}$ has measure zero.
  - Given $\lambda > 0$, $C > 0$ and $d \in \{1, 2, ...\}$, the family of mappings $f : \mathbb{X} \to \mathbb{Y}$ such that

$$(9.3) \qquad \int_\mathbb{X} e^{\lambda K(x)} \, dx \leqslant C$$

$$(9.4) \qquad \deg(f; \mathbb{X}, \mathbb{Y}) \leqslant d$$

is compact with respect to uniform convergence.

- If for sufficiently large $\lambda = \lambda(n)$
$$\int_{\mathbb{X}} e^{\lambda K(x)}\, dx < \infty$$
then $f \in \mathscr{W}^{1,n}(\mathbb{X}, \mathbb{Y})$.

We shall not prove this theorem, it can be found in [28], [29], [30], [31], [38], [39], [40], and [41].

Acknowledgements

Part of this research was done while Onninen was visiting Mathematics Department at Syracuse University. He wishes to thank SU for the support and hospitality. The research was partially carried out during Hajłasz stay in the Department of Mathematics at University of Michigan. He wishes to thank UM for the support and hospitality. Also Malý was present in the Department of Mathematics at UM at the very key moment. We wish to thank Juha Heinonen, Pekka Koskela and UM for arranging and supporting this short but important visit. We are grateful to Vladimír Šverák and Stefan Müller for stimulating discussions and pointing out the question concerning Hardy spaces.

Bibliography

[1] J. M. Ball, *Convexity conditions and existence theorems in nonlinear elasticity.* Arch. Rational Mech. Anal. **63** (1976/77), 337-403.
[2] F. Bethuel, *The approximation problem for Sobolev maps between two manifolds.* Acta Math. **167** (1991), 153-206.
[3] F. Bethuel and X. Zheng, *Density of smooth functions between two manifolds in Sobolev spaces.* J. Funct. Anal. **80** (1988), 60-75.
[4] J. Bourgain, H. Brezis and P. Mironescu, *Lifting in Sobolev spaces.* J. Anal. **80** (2000), 37-86.
[5] H. Brezis, Y. Li, P. Mironescu and L. Nirenberg, *Degree and Sobolev spaces.* Topol. Methods Nonlinear Anal. **13** (1999), 181-190.
[6] H. Brezis and L. Nirenberg, *Degree theory and BMO, Part I: Compact manifolds without boundaries.* Selecta Math. **1** (1995), 197-263.
[7] H. Brezis, and L. Nirenberg, *Degree theory and BMO. II. Compact manifolds with boundaries.* Selecta Math. (N.S.) **2** (1996), 309-368.
[8] H. Cartan, *Differential forms,* Houghton Mifflin Co., Boston, 1970.
[9] R. Coifman, P. L. Lions, Y. Meyer, and S. Semmes, *Compensated compactness and Hardy spaces.* J. Math. Pures Appl. (9) **72** (1993), 247-286.
[10] J. Eells and L. Lemaire, (1978). *A report on harmonic maps.* Bull. London Math. Soc. **10** (1978), 1-68.
[11] M. J. Esteban and S. Müller, *Sobolev maps with integer degree and applications to Skyrme's problem.* Proc. Roy. Soc. London Ser. A **436** (1992), no. 1896, 197-201.
[12] H. Federer, *Geometric measure theory,* Springer, 1969.
[13] F. Giannetti, T. Iwaniec, J. Onninen, and A. Verde,(2002). *Estimates of Jacobians by subdeterminants.* J. Geom. Anal. **12** (2002), 223-254.
[14] L. Greco, T. Iwaniec, C. Sbordone and B. Stroffolini, *Degree formulas for maps with nonintegrable Jacobian.* Topol. Methods Nonlinear Anal. **6** (1995), 81-95.
[15] M. J. Greenberg and J. R. Harper, *Algebraic topology. A first course,* Mathematics Lecture Note Series, vol. 58, Benjamin/Cummings Publishing Co., Inc., Advanced Book Program, Reading, Mass. 1981.
[16] M. L. Gromov and V. A. Rohlin, *Imbeddings and immersions in Riemannian geometry.* (Russian) Uspehi Mat. Nauk **25** (1970), no. 5 (155), 3-62.
[17] P. Hajłasz, *Approximation of Sobolev mappings.* Nonlinear Anal. **22** (1994), 1579-1591.
[18] P. Hajłasz, *Equivalent statement of the Poincaré conjecture.* Ann. Mat. Pura Appl. (4) **167** (1994), 25-31.
[19] P. Hajłasz, *Sobolev Mappings: Lipschitz density is not a bi-Lipschitz invariant of the target.* Geom. Funct. Anal. **17** (2007), 435–467.
[20] P. Hajłasz, *Density of Lipschitz mappings in the class of Sobolev mappings between metric spaces.* Math. Ann. (to appear).
[21] P. Hajłasz and J. Malý, *Approximation in Sobolev spaces of nonlinear expressions involving the gradient.* Ark. Mat. **40** (2002), 245–274.
[22] C. Hamburger, *Some properties of the degree for a class of Sobolev maps.* Proc. Roy. Soc. London Ser. A **455** (1999), no. 1986, 2331-2349.
[23] F. Hang, and F. Lin, *Topology of Sobolev mappings.* Math. Res. Lett. **8** (2001), 321-330.
[24] F. Hang and F. Lin, *Topology of Sobolev mappings II.* Acta Math. **191** (2003), 55–107.
[25] J. Heinonen, T. Kilpeläinen and O. Martio, *Nonlinear potential theory of degenerate elliptic equations,* Oxford Mathematical Monographs, 1993.
[26] F. Hélein, *Harmonic maps, conservation laws and moving frames,* Second edition. Cambridge Tracts in Mathematics, vol. 150. Cambridge University Press, Cambridge, 2002.
[27] T. Iwaniec, *Nonlinear differential forms,* Lectures in Jyväskylä. Report No. 80. University of Jyväskylä, Department of Mathematics, Jyväskylä, 1998.

[28] T. Iwaniec, P. Koskela and G. J. Martin, *Mappings of BMO-distortion and Beltrami-type operators.* J. Anal. Math. **88** (2002), 337–381.

[29] T. Iwaniec, P. Koskela, G. Martin and C. Sbordone, *Mappings of finite distortion: $L^n \log^\chi L$-integrability.* J. London Math. Soc. (2) **67** (2003), 123–136.

[30] T. Iwaniec, P. Koskela and J. Onninen, *Mappings of finite distortion: Monotonicity and continuity.* Invent. Math. **144** (2001), 507-531.

[31] T. Iwaniec, P. Koskela and J. Onninen, *Mappings of finite distortion: compactness.* Ann. Acad. Sci. Fenn. Math. 27 (2002), 391–417.

[32] T. Iwaniec and G. J. Martin, *Geometric Function Theory and Non-linear Analysis,* Oxford Mathematical Monographs, 2001.

[33] T. Iwaniec and J. Onninen, *$\mathcal{H}^1$-Estimates of Jacobians by Subdeterminants.* Math. Ann. **324** (2002), 341-358.

[34] T. Iwaniec and C. Sbordone, *On the integrability of the Jacobian under minimal hypotheses.* Arch. Rational Mech. Anal. **119** (1992), 129-143.

[35] T. Iwaniec, C. Scott and B. Stroffolini, *Nonlinear Hodge theory on manifolds with boundary.* Ann. Mat. Pura Appl. (4) **177** (1999), 37-115.

[36] T. Iwaniec and A. Verde, *A study of Jacobians in Hardy-Orlicz spaces.* Proc. Roy. Soc. Edinburgh Sect. A **129** (1999), 539-570.

[37] J. Jost, *Riemannian geometry and geometric analysis,* Third edition. Universitext. Springer-Verlag, Berlin, 2002.

[38] J. Kauhanen, P. Koskela and J. Malý, *Mappings of finite distortion: Discreteness and openness.* Arch. Rational Mech. Anal. **160** (2001), 135-151.

[39] J. Kauhanen, P. Koskela and J. Malý, *Mappings of finite distortion: Condition N.* Michigan Math. J. **49** (2001), 169–181.

[40] J. Kauhanen, P. Koskela, J. Malý, J. Onninen and X. Zhong, *Mappings of finite distortion: Sharp Orlicz-conditions.* Rev. Mat. Iberoamericana **17** (2003), 857–872.

[41] P. Koskela and J. Malý, (2003). *Mappings of finite distortion: The zero set of the Jacobian.* J. Eur. Math. Soc. (JEMS) **5** (2003), 95–105.

[42] P. Koskela and X. Zhong, *Minimal assumptions for the integrability of the Jacobian.* Ricerche Mat. **51** (2003), 297–311.

[43] C. B. Morrey, *Multiple Integrals in the Calculus of Variations,* Springer-Verlag, Berlin, 1966.

[44] S. Müller, *Higher integrability of determinants and weak convergence in L^1.* J. Reine Angew. Math. **412** (1990), 20–34.

[45] S. Müller, *Det = det. A remark on the distributional determinant.* C. R. Acad. Sci. Paris Sr. I Math. **311** (1990), 13–17.

[46] S. Müller and V. Šverák, *On surfaces of finite total curvature.* J. Differential Geom. **42** (1995), 229-258.

[47] J. Nash, *The imbedding problem for Riemannian manifolds.* Ann. of Math. (2) **63** (1956), 20-63.

[48] L. Saloff-Coste, *Aspects of Sobolev-type inequalities,* London Mathematical Society Lecture Note Series, vol. 289. Cambridge University Press, Cambridge, 2002.

[49] R. Schoen and K. Uhlenbeck, *A regularity theory for harmonic maps.* J. Differential Geom. **17** (1982), 307-335.

[50] R. Schoen and K. Uhlenbeck, *Boundary regularity and the Dirichlet problem for harmonic maps.* J. Differential Geom. **18** (1983), 253–268.

[51] C. Scott, *L^p-theory of differential forms on manifolds.* Trans. Amer. Math. Soc., **347** (1995), 2075-2096.

[52] E. M. Stein, *Note on the class $L \log L$.* Studia Math. **32** (1969), 305-310.

[53] E. M. Stein, *Harmonic analysis: real-variable methods, orthogonality, and oscillatory integrals.* Princeton Mathematical Series, 43. Monographs in Harmonic Analysis, III. Princeton University Press, Princeton, NJ, 1993.

[54] R. S. Strichartz, *The Hardy space H^1 on manifolds and submanifolds.* Canad. J. Math. **24** (1972), 915-925.

[55] F. W. Warner, *Foundations of differentiable manifolds and Lie groups,* Graduate Texts in Mathematics, 94. Springer-Verlag, New York-Berlin, 1983.

[56] B. White, *Infima of energy functionals in homotopy classes of mappings.* J. Differential Geom. **23** (1986), 127-142.

[57] B. White, *Homotopy classes in Sobolev spaces and the existence of energy minimizing maps.* Acta Math. **160** (1988), 1-17.

Editorial Information

To be published in the *Memoirs*, a paper must be correct, new, nontrivial, and significant. Further, it must be well written and of interest to a substantial number of mathematicians. Piecemeal results, such as an inconclusive step toward an unproved major theorem or a minor variation on a known result, are in general not acceptable for publication.

Papers appearing in *Memoirs* are generally at least 80 and not more than 200 published pages in length. Papers less than 80 or more than 200 published pages require the approval of the Managing Editor of the Transactions/Memoirs Editorial Board.

As of November 30, 2007, the backlog for this journal was approximately 14 volumes. This estimate is the result of dividing the number of manuscripts for this journal in the Providence office that have not yet gone to the printer on the above date by the average number of monographs per volume over the previous twelve months, reduced by the number of volumes published in four months (the time necessary for preparing a volume for the printer). (There are 6 volumes per year, each usually containing at least 4 numbers.)

A Consent to Publish and Copyright Agreement is required before a paper will be published in the *Memoirs*. After a paper is accepted for publication, the Providence office will send a Consent to Publish and Copyright Agreement to all authors of the paper. By submitting a paper to the *Memoirs*, authors certify that the results have not been submitted to nor are they under consideration for publication by another journal, conference proceedings, or similar publication.

Information for Authors

Memoirs are printed from camera copy fully prepared by the author. This means that the finished book will look exactly like the copy submitted.

Initial submission. The AMS uses Centralized Manuscript Processing for initial submissions. Authors should submit a PDF file using the Initial Manuscript Submission form found at www.ams.org/cgi-bin/peertrack/submission.pl, or send one copy of the manuscript to the following address: Centralized Manuscript Processing, MEMOIRS OF THE AMS, 201 Charles Street, Providence, RI 02904-2294 USA. If a paper copy is being forwarded to the AMS, indicate that it is for it Memoirs and include the name of the corresponding author, contact information such as email address or mailing address, and the name of an appropriate Editor to review the paper (see the list of Editors below).

The paper must contain a *descriptive title* and an *abstract* that summarizes the article in language suitable for workers in the general field (algebra, analysis, etc.). The *descriptive title* should be short, but informative; useless or vague phrases such as "some remarks about" or "concerning" should be avoided. The *abstract* should be at least one complete sentence, and at most 300 words. Included with the footnotes to the paper should be the 2000 *Mathematics Subject Classification* representing the primary and secondary subjects of the article. The classifications are accessible from www.ams.org/msc/. The list of classifications is also available in print starting with the 1999 annual index of *Mathematical Reviews*. The Mathematics Subject Classification footnote may be followed by a list of *key words and phrases* describing the subject matter of the article and taken from it. Journal abbreviations used in bibliographies are listed in the latest *Mathematical Reviews* annual index. The series abbreviations are also accessible from www.ams.org/publications/. To help in preparing and verifying references, the AMS offers MR Lookup, a Reference Tool for Linking, at www.ams.org/mrlookup/.

Electronically prepared manuscripts. The AMS encourages electronically prepared manuscripts, with a strong preference for $\mathcal{AMS}$-LaTeX. To this end, the Society has prepared $\mathcal{AMS}$-LaTeX author packages for each AMS publication. Author packages include instructions for preparing electronic manuscripts, samples, and a style file that generates the particular design specifications of that publication series. Though $\mathcal{AMS}$-LaTeX is the highly preferred format of TeX, author packages are also available in $\mathcal{AMS}$-TeX.

Authors may retrieve an author package from the AMS website starting from www.ams.org/tex/ or via FTP to ftp.ams.org (login as anonymous, enter username as password, and type cd pub/author-info). The *AMS Author Handbook* and the *Instruction Manual* are available in PDF format following the author packages link from www.ams.org/tex/. The author package can also be obtained free of charge by sending email to tech-support@ams.org (Internet) or from the Publication Division, American Mathematical Society, 201 Charles St., Providence, RI 02904-2294, USA. When requesting an author package, please specify $\mathcal{AMS}$-LaTeX or $\mathcal{AMS}$-TeX and the publication in which your paper will appear. Please be sure to include your complete mailing address.

After acceptance. The final version of the electronic file should be sent to the Providence office (this includes any TeX source file, any graphics files, and the DVI or PostScript file) immediately after the paper has been accepted for publication.

Before sending the source file, be sure you have proofread your paper carefully. The files you send must be the EXACT files used to generate the proof copy that was accepted for publication. For all publications, authors are required to send a printed copy of their paper, which exactly matches the copy approved for publication, along with any graphics that will appear in the paper.

Accepted electronically prepared files can be submitted via the web at www.ams.org/submit-book-journal/, sent via FTP, or sent on CD-Rom or diskette to the Electronic Prepress Department, American Mathematical Society, 201 Charles Street, Providence, RI 02904-2294 USA. TeX source files, DVI files, and PostScript files can be transferred over the Internet by FTP to the Internet node ftp.ams.org (130.44.1.100). When sending a manuscript electronically via CD-Rom or diskette, please be sure to include a message identifying the paper as a Memoir.

Electronically prepared manuscripts can also be sent via email to pub-submit@ams.org (Internet). In order to send files via email, they must be encoded properly. (DVI files are binary and PostScript files tend to be very large.)

Electronic graphics. Comprehensive instructions on preparing graphics are available at www.ams.org/jourhtml/. A few of the major requirements are given here.

Submit files for graphics as EPS (Encapsulated PostScript) files. This includes graphics originated via a graphics application as well as scanned photographs or other computer-generated images. If this is not possible, TIFF files are acceptable as long as they can be opened in Adobe Photoshop or Illustrator. No matter what method was used to produce the graphic, it is necessary to provide a paper copy to the AMS.

Authors using graphics packages for the creation of electronic art should also avoid the use of any lines thinner than 0.5 points in width. Many graphics packages allow the user to specify a "hairline" for a very thin line. Hairlines often look acceptable when proofed on a typical laser printer. However, when produced on a high-resolution laser imagesetter, hairlines become nearly invisible and will be lost entirely in the final printing process.

Screens should be set to values between 15% and 85%. Screens which fall outside of this range are too light or too dark to print correctly. Variations of screens within a graphic should be no less than 10%.

Inquiries. Any inquiries concerning a paper that has been accepted for publication should be sent to memo-query@ams.org or directly to the Electronic Prepress Department, American Mathematical Society, 201 Charles St., Providence, RI 02904-2294 USA.

Editors

This journal is designed particularly for long research papers, normally at least 80 pages in length, and groups of cognate papers in pure and applied mathematics. Papers intended for publication in the *Memoirs* should be addressed to one of the following editors. The AMS uses Centralized Manuscript Processing for initial submissions to AMS journals. Authors should follow instructions listed on the Initial Submission page found at www.ams.org/memo/memosubmit.html.

Algebra to ALEXANDER KLESHCHEV, Department of Mathematics, University of Oregon, Eugene, OR 97403-1222; email: ams@noether.uoregon.edu

Algebraic geometry and its application to MINA TEICHER, Emmy Noether Research Institute for Mathematics, Bar-Ilan University, Ramat-Gan 52900, Israel; email: teicher@macs.biu.ac.il

Algebraic geometry to DAN ABRAMOVICH, Department of Mathematics, Brown University, Box 1917, Providence, RI 02912; email: amsedit@math.brown.edu

Algebraic number theory to V. KUMAR MURTY, Department of Mathematics, University of Toronto, 100 St. George Street, Toronto, ON M5S 1A1, Canada; email: murty@math.toronto.edu

Algebraic topology to ALEJANDRO ADEM, Department of Mathematics, University of British Columbia, Room 121, 1984 Mathematics Road, Vancouver, British Columbia, Canada V6T 1Z2; email: adem@math.ubc.ca

Combinatorics to JOHN R. STEMBRIDGE, Department of Mathematics, University of Michigan, Ann Arbor, Michigan 48109-1109; email: FRS@umich.edu

Complex analysis and harmonic analysis to ALEXANDER NAGEL, Department of Mathematics, University of Wisconsin, 480 Lincoln Drive, Madison, WI 53706-1313; email: nagel@math.wisc.edu

Differential geometry and global analysis to LISA C. JEFFREY, Department of Mathematics, University of Toronto, 100 St. George St., Toronto, ON Canada M5S 3G3; email: jeffrey@math.toronto.edu

Functional analysis and operator algebras to DIMITRI SHLYAKHTENKO, Department of Mathematics, University of California, Los Angeles, CA 90095; email: shlyakht@math.ucla.edu

Geometric analysis to WILLIAM P. MINICOZZI II, Department of Mathematics, Johns Hopkins University, 3400 N. Charles St., Baltimore, MD 21218; email: trans@math.jhu.edu

Geometric analysis to MARK FEIGHN, Math Department, Rutgers University, Newark, NJ 07102; email: feighn@andromeda.rutgers.edu

Harmonic analysis, representation theory, and Lie theory to ROBERT J. STANTON, Department of Mathematics, The Ohio State University, 231 West 18th Avenue, Columbus, OH 43210-1174; email: stanton@math.ohio-state.edu

Logic to STEFFEN LEMPP, Department of Mathematics, University of Wisconsin, 480 Lincoln Drive, Madison, Wisconsin 53706-1388; email: lempp@math.wisc.edu

Number theory to JONATHAN ROGAWSKI, Department of Mathematics, University of California, Los Angeles, CA 90095; email: jonr@math.ucla.edu

Partial differential equations to GUSTAVO PONCE, Department of Mathematics, South Hall, Room 6607, University of California, Santa Barbara, CA 93106; email: ponce@math.ucsb.edu

Partial differential equations and dynamical systems to PETER POLACIK, School of Mathematics, University of Minnesota, Minneapolis, MN 55455; email: polacik@math.umn.edu

Probability and statistics to RICHARD BASS, Department of Mathematics, University of Connecticut, Storrs, CT 06269-3009; email: bass@math.uconn.edu

Real analysis and partial differential equations to DANIEL TATARU, Department of Mathematics, University of California, Berkeley, Berkeley, CA 94720; email: tataru@math.berkeley.edu

All other communications to the editors should be addressed to the Managing Editor, ROBERT GURALNICK, Department of Mathematics, University of Southern California, Los Angeles, CA 90089-1113; email: guralnic@math.usc.edu.

Titles in This Series

900 **Wolfgang Bertram,** Differential geometry, Lie groups and symmetric spaces over general base fields and rings, 2008

899 **Piotr Hajłasz, Tadeusz Iwaniec, Jan Malý, and Jani Onninen,** Weakly differentiable mappings between manifolds, 2008

898 **John Rognes,** Galois extensions of structured ring spectra/Stably dualizable groups, 2008

897 **Michael I. Ganzburg,** Limit theorems of polynomial approximation with exponential weights, 2008

896 **Michael Kapovich, Bernhard Leeb, and John J. Millson,** The generalized triangle inequalities in symmetric spaces and buildings with applications to algebra, 2008

895 **Steffen Roch,** Finite sections of band-dominated operators, 2008

894 **Martin Dindoš,** Hardy spaces and potential theory on C^1 domains in Riemannian manifolds, 2008

893 **Tadeusz Iwaniec and Gaven Martin,** The Beltrami Equation, 2008

892 **Jim Agler, John Harland, and Benjamin J. Raphael,** Classical function theory, operator dilation theory, and machine computation on multiply-connected domains, 2008

891 **John H. Hubbard and Peter Papadopol,** Newton's method applied to two quadratic equations in $\mathbb{C}^2$ viewed as a global dynamical system, 2008

890 **Steven Dale Cutkosky,** Toroidalization of dominant morphisms of 3-folds, 2007

889 **Michael Sever,** Distribution solutions of nonlinear systems of conservation laws, 2007

888 **Roger Chalkley,** Basic global relative invariants for nonlinear differential equations, 2007

887 **Charlotte Wahl,** Noncommutative Maslov index and eta-forms, 2007

886 **Robert M. Guralnick and John Shareshian,** Symmetric and alternating groups as monodromy groups of Riemann surfaces I: Generic covers and covers with many branch points, 2007

885 **Jae Choon Cha,** The structure of the rational concordance group of knots, 2007

884 **Dan Haran, Moshe Jarden, and Florian Pop,** Projective group structures as absolute Galois structures with block approximation, 2007

883 **Apostolos Beligiannis and Idun Reiten,** Homological and homotopical aspects of torsion theories, 2007

882 **Lars Inge Hedberg and Yuri Netrusov,** An axiomatic approach to function spaces, spec tral synthesis and Luzin approximation, 2007

881 **Tao Mei,** Operator valued Hardy spaces, 2007

880 **Bruce C. Berndt, Geumlan Choi, Youn-Seo Choi, Heekyoung Hahn, Boon Pin Yeap, Ae Ja Yee, Hamza Yesilyurt, and Jinhee Yi,** Ramanujan's forty identities for Rogers-Ramanujan functions, 2007

879 **O. García-Prada, P. B. Gothen, and V. Muñoz,** Betti numbers of the moduli space of rank 3 parabolic Higgs bundles, 2007

878 **Alessandra Celletti and Luigi Chierchia,** KAM stability and celestial mechanics, 2007

877 **María J. Carro, José A. Raposo, and Javier Soria,** Recent developments in the theory of Lorentz spaces and weighted inequalities, 2007

876 **Gabriel Debs and Jean Saint Raymond,** Borel liftings of Borel sets: Some decidable and undecidable statements, 2007

875 **C. Krattenthaler and T. Rivoal,** Hypergéométrie et fonction zêta de Riemann, 2007

874 **Sonia Natale,** Semisolvability of semisimple Hopf algebras of low dimension, 2007

873 **A. J. Duncan,** Exponential genus problems in one-relator products of groups, 2007

872 **Anthony V. Geramita, Tadahito Harima, Juan C. Migliore, and Yong Su Shin,** The Hilbert function of a level algebra, 2007

871 **Pascal Auscher,** On necessary and sufficient conditions for L^p-estimates of Riesz transforms associated to elliptic operators on $\mathbb{R}^n$ and related estimates, 2007

TITLES IN THIS SERIES

- 870 **Takuro Mochizuki,** Asymptotic behaviour of tame harmonic bundles and an application to pure twistor D-modules, Part 2, 2007
- 869 **Takuro Mochizuki,** Asymptotic behaviour of tame harmonic bundles and an application to pure twistor D-modules, Part 1, 2007
- 868 **Gelu Popescu,** Entropy and multivariable interpolation, 2006
- 867 **Vilmos Totik,** Metric properties of harmonic measures, 2006
- 866 **William Craig,** Semigroups underlying first-order logic, 2006
- 865 **Nathanial P. Brown,** Invariant means and finite representation theory of $C*$-algebras, 2006
- 864 **John M. Lee,** Fredholm operators and Einstein metrics on conformally compact manifolds, 2006
- 863 **M. Lübke and A. Teleman,** The Universal Kobayashi-Hitchin correspondence on Hermitian manifolds, 2006
- 862 **Alberto Canonaco,** The Beilinson complex and canonical rings of irregular surfaces, 2006
- 861 **Leon A. Takhtajan and Lee-Peng Teo,** Weil-Petersson metric on the universal Teichmüller space, 2006
- 860 **Thomas M. Fiore,** Pseudo limits, biadjoints and pseudo algebras: Categorical foundations of conformal field theory, 2006
- 859 **N. Arcozzi, R. Rochberg, and E. Sawyer,** Carleson measures and interpolating sequences for Besov spaces on complex balls, 2006
- 858 **Enrico Valdinoci, Berardino Sciunzi, and Vasile Ovidiu Savin,** Flat level set regularity of p-Laplace phase transitions, 2006
- 857 **Donatella Danielli, Nocola Garofalo, and Duy-Minh Nhieu,** Non-doubling Ahlfors measures, perimeter measures, and the characterization of the trace spaces of Sobolev functions in Carnot-Carathéodory spaces, 2006
- 856 **Vladimir Bolotnikov and Harry Dym,** On boundary interpolation for matrix valued Schur functions, 2006
- 855 **Yevgenia Kashina, Yorck Sommerhäuser, and Yongchang Zhu,** On higher Frobenius-Schur indicators, 2006
- 854 **Noam Greenberg,** The role of true finiteness in the admissible recursively enumerable degrees, 2006
- 853 **Joachim Krieger,** Stability of spherically symmetric wave maps, 2006
- 852 **Viorel Barbu, Irena Lasiecka, and Roberto Triggiani,** Tangential boundary stabilization of Navier-Stokes equations, 2006
- 851 **Jie Wu,** On maps from loop suspensions to loop spaces and the shuffle relations on the Cohen groups, 2006
- 850 **Siegfried Echterhoff, S. Kaliszewski, John Quigg, and Iain Raeburn,** A categorical approach to imprimitivity theorems for C^*-dynamical systems, 2006
- 849 **Katsuhiko Kuribayashi, Mamoru Mimura, and Tetsu Nishimoto,** Twisted tensor products related to the cohomology of the classifying spaces of loop groups, 2006
- 848 **Bob Oliver,** Equivalences of classifying spaces completed at the prime two, 2006
- 847 **Eric T. Sawyer and Richard L. Wheeden,** Hölder continuity of weak solutions to subelliptic equations with rough coefficients, 2006
- 846 **Victor Beresnevich, Detta Dickinson, and Sanju Velani,** Measure theoretic laws for lim–sup sets, 2006

For a complete list of titles in this series, visit the
AMS Bookstore at **www.ams.org/bookstore/**.